The Physics of the Deformation of Densely Packed Granular Materials

The Physics of the Deformation of Densely Packed Granular Materials

M.A.C. Koenders

World Scientific

NEW JERSEY • LONDON • SINGAPORE • BEIJING • SHANGHAI • HONG KONG • TAIPEI • CHENNAI • TOKYO

Published by

World Scientific Publishing Europe Ltd.

57 Shelton Street, Covent Garden, London WC2H 9HE

Head office: 5 Toh Tuck Link, Singapore 596224

USA office: 27 Warren Street, Suite 401-402, Hackensack, NJ 07601

Library of Congress Cataloging-in-Publication Data
Names: Koenders, M. A. C., author.
Title: The physics of the deformation of densely packed granular materials / M.A.C. Koenders.
Description: New Jersey : World Scientific, [2020] | Includes bibliographical references and index.
Identifiers: LCCN 2019053417 | ISBN 9781786348234 (hardcover) | ISBN 9781786348241 (ebook) |
ISBN 9781786348258 (ebook other)
Subjects: LCSH: Granular materials. | Deformations (Mechanics)--Mathematical models.
Classification: LCC TA418.78 .K63 2020 | DDC 620/.43--dc23
LC record available at https://lccn.loc.gov/2019053417

British Library Cataloguing-in-Publication Data
A catalogue record for this book is available from the British Library.

For any available supplementary material, please visit
https://www.worldscientific.com/worldscibooks/10.1142/Q0244#t=suppl

Desk Editors: Herbert Moses/Shi Ying Koe

Typeset by Stallion Press
Email: enquiries@stallionpress.com

Printed in Singapore

Preface

In the early 1980s, the idea first took hold that the mechanical response of a dense granular medium can be understood from a basis of the inter-particle contact properties. The initial efforts, a mean-field theory, had very poor results and papers on 'micro-mechanics' were usually relegated to the last section of conference proceedings. Gradually, the insight came about that a granular medium cannot be captured in a mean-field theory and that some form of non-homogeneity in the fabric properties has to be accounted for. The beginnings of this concept were implemented using the available continuum theories on heterogeneity and a few papers came out in the early 90s showing that in certain special cases the mechanical response was captured, but — irritatingly — not all cases. Highly anisotropic packed beds, for example, could not be accounted for and the failure of a granular medium at high stress ratio remained a mystery.

While progress since then has been slow, it is now clear that a proper theory of granular deformation *must* include a method that deals with heterogeneity that is particularly applicable to a system of particles. This turns out to be the theory of 'connected media', which captures the physics of contacting particulates in an appropriate manner. It has also been extended to anisotropic cases. A rigorous approach to Coulomb friction as an inter-particle interaction is required as well. Together these developments can now be implemented with great success.

To preserve analytical insight it is advantageous to use simplified models with round particles and on occasion do a two-dimensional calculation, rather than a three-dimensional one. This does not matter for the understanding of the physics that is at play. The theories have also been applied successfully to other fields where the inter-particle interaction has a more chemical character. Filter cake formation (relevant to chemical engineering) is an example. Due to the large number of natural occurrences and applications of dense granular matter there is relevance in a variety of disciplines.

This book presents a detailed exposition of all the concepts and mathematical techniques that are necessary to understand the current state of the subject. The student from a non-mathematical background may initially have to put in a certain amount of work to grasp the intricacies of the line of argument. This is a very algebraic subject; there is not much one can do about that. However, a mathematical appendix and an introductory chapter on continuum mechanics and Cartesian tensor calculus are provided to make the journey easier.

Curt Koenders
Canterbury, 2019

About the Author

M.A.C. Koenders is a physicist who has worked in industry, academia and as a consultant. His expertise is in the mechanics of granular media. He has collaborated with physicists, civil engineers, chemical engineers, mathematicians and geologists. He has some 200 (co)-authored papers in refereed journals, book chapters and conference proceedings relevant to these subjects. He has also been a substantial fund-raiser to benefit the progress of the subject and collaborated on many projects to further the understanding of aspects of granular mechanics.

In industry he was active in drawing up geotechnical filtration rules for the soil mechanics community. In other work on civil engineering he has had a long-standing collaboration with the Bundesanstalt für Wasserbau, Karlsruhe to provide a theory on the mechanics of unsaturated soils.

In chemical engineering he collaborated to describe and improve filtration processes, especially the mechanics of cake formation. He was awarded the gold medal of the Filtration Society (2003) for his contributions to oscillated filtration.

In geology he contributed to the understanding of processes under volcanoes, especially co-authoring papers on magma infiltration into dilatant granular layers.

In mathematics he introduced the concepts of structures formation in granular materials and non-Newtonian flow processes through granular masses.

He has supervised 13 PhD students. He is a member of the Institute of Physics and currently associated with the University of Southampton.

Contents

Chapter 1

General Concepts

1.1 Introduction

Granular materials play a role in nearly all human activities. Users of, for example, sand, from children in sandpits to sophisticated geotechnical engineers, know that it is a fascinating — and to some extent, unpredictable — material. Many groups are concerned professionally with granular materials: chemical engineers, pharmacists, food technologists, agriculturalists, biologists, geologists, geophysicists, astronomers even, are obliged to study their behaviour under a wide variety of circumstances. In addition to sand, which itself may be of many compositions, the types of materials include gravel, fine-particle aggregates as employed in cosmetics, pharmaceuticals, dust, crushed rock and granules that occur in a domestic environment, such as breakfast cereals, sugar, salt and (instant or ground) coffee granules.

It is important to distinguish between the various states in which these materials may be encountered. The possible range of regimes is extensive. The delineation of regimes is accomplished by specifying first the packing density, second the grain-size, or size distribution and particle shape, then the type of medium in the interstices between the grains (fluid, gas, vacuum) and finally the stress and temperature environment. Depending on any combination of these factors, examples of phenomena that may take place come in a wide variety. A few celebrated ones are landslides, blocked silos and sewers, rubble asteroids breaking up, segregation effects in breakfast cereals, dust-storm propagation after a terrorist attack, the spreading of sun-cream over skin and the formation of dunes. The list is by no means exhaustive; not only do people continually invent new applications for granulates, they also discover new processes where these materials may be deployed. The sheer diversity of effects illustrates the range of professionals that may be engaged with the subject.

The mechanical behaviour of an assembly of grains depends first and foremost on the interaction between the particles. For a low packing density the grains are fairly free to move and interactions may take place in a similar way to the molecules in a gas: the interactions are short-duration 'events'. When, on the other hand, the material is densely packed the grains are locked in enduring interaction with each other. This does not preclude relative motion between the particles. In a dense slurry, for example, the interstitial fluid is the interactive medium. Particles may move, while the interactive strength varies with motion, but the interaction continues to be relevant for particles in close proximity. When the medium is dense and dry, on the other hand, particles must make contact. Their relative motion may be sliding, or even suffer a very slight indentation when two particles are being pressed together hard, but there is only a non-zero interaction while the contact endures.

In order to describe the motion in various states, distinctly different branches of mechanics are required. For a dilute flow in which collisions are prevalent, for example, concepts of gas dynamics have to be invoked: a temperature field is needed to describe the velocity fluctuations while motion takes place. For dense (but not too dense) slurry flow in which a fluid mediates the interaction of the grains the relative velocity difference of the particles needs to be described. For very small particles Brownian motion will play a role too. For a dense packing, in which the grains are in enduring contact, the physics of the interaction is quite different. As this is the field of interest in the publication to hand a small study of the background of this subject is of use.

It could be argued that the densely packed state is fairly boring, as the displacements tend to be so insignificant. Essentially, one might say, a densely packed granular material behaves like a solid. There are, however, certain features that relate to this régime that are quite unlike traditional solids. In fact, it is one of the most difficult to describe problems in materials science. The reason for this is that the material properties change dramatically under certain specific loading conditions.

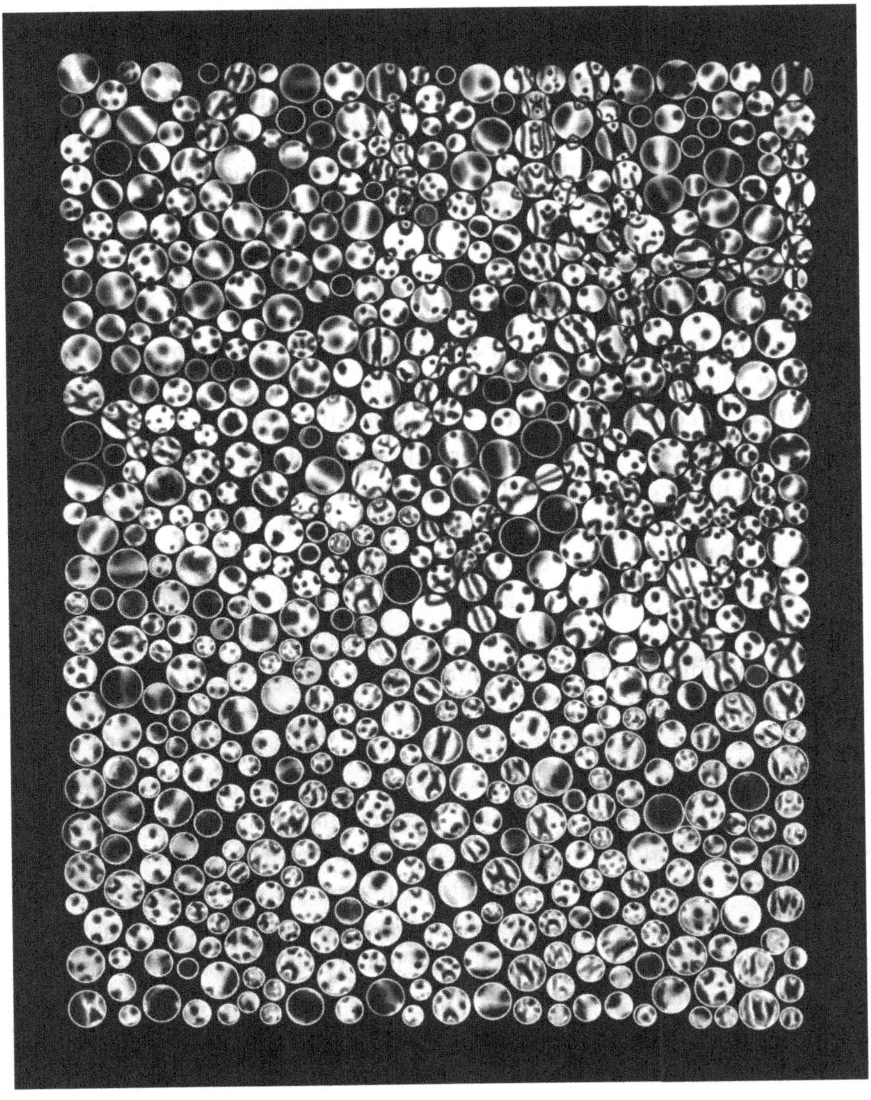

Figure 1.1. Picture of an assembly of photo-elastic discs. Experiment by [Konishi, 1978].

The easiest way to see where the problems arise is by considering an experiment of dry dense sand on a slope. When the sand is initially deposited and densified the slope is horizontal. Now imagine an experiment in which the slope angle is gradually increased. There comes a point when the angle is so great that the sand can no longer support a stable configuration and a landslide ensues. The changes in the sand up until this point are almost imperceptible, yet, internally, changes must have taken place in order for the sand to go into a state that cannot support a stable equilibrium. The question is: what physics underlies the internal change of state and how can its mechanics be captured?

This problem is, of course, the province of soil mechanics. Tribute must be paid to the tremendous body of useful work that has been produced by civil engineers, especially in the area of experimentation. One type of test in particular is very common in soil testing and that is the so-called *triaxial-cell test*. In this test a cylindrical sample of soil is subjected to a stress path in which — after initially building up a compressive pressure — the stress on the cylinder wall is kept constant while the stress on the ends of the cylinder is increased (precise definitions of stress and strain are explored in Chapter 2). The same type of test can be done in two dimensions on a sample of an assembly of discs. The latter case is illustrated in a picture of photo-elastic discs in which the contact forces are made visible by means of polarised light. Figure 1.1 provides an example.

A typical response of the assembly so stressed is depicted below in somewhat stylised form (stylised to remove the inevitable experimental noise). The stress ratio (the ratio of the major and minor principal stress) goes up with increased principal strain until it appears to remain more or less constant. Now look at the tangent modulus (ratio of stress increment to strain increment). While the stress ratio is close to unity the assembly is quite stiff and behaves just like a solid block of material. As the stress ratio increases, however, the tangent modulus rapidly decreases till it reaches zero — a dramatic change in only a few percent of deformation!

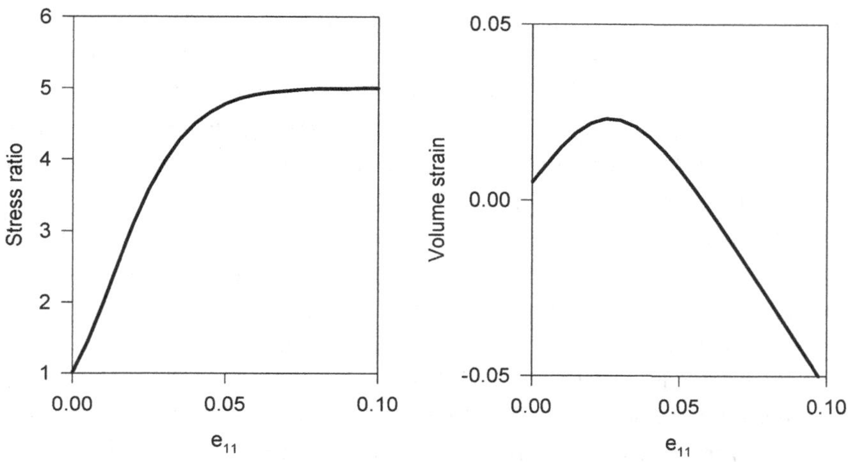

Figure 1.2. Stress ratio and volume strain as a function of the major principal strain in a biaxial cell test.

Even more bizarre is the behaviour of the volume strain. Initially, at a stress ratio close to unity, the sample contracts, as one would expect from an ordinary solid that is compressed in one direction. At higher stress ratios a peculiar effect becomes manifest: the sample expands. This is completely counter-intuitive behaviour. The effect is called *dilation*. The reader may carry out a very simple experiment to verify the phenomenon. Go to a wet beach with well-compacted sand and simply step on it. One can see the sand go dry underneath one's feet. The soil expands, causing there to be more space in the interstices, and in so doing it sucks the water in from the neighbourhood, making it temporarily drier. The effect was first described by [Reynolds, 1885].

The amount of motion involved in this development is minimal; the strain is in the order of a few percent. The mechanical features that occur here are very important not only for the geotechnical industry, but also for the understanding of, for example, the motion of burrowing animals — see for example, [Dorgan *et al.*, 2006]. While a further discussion is only possible when grain assemblies are considered that contain an interstitial fluid, it is obvious that such creatures are adapted to employ the mechanical properties of granular deposits, such as dilatancy, in their survival. Another example is the motion of sheared layers of granular materials in geological settings — see [Petford and Koenders, 2003] — in which hot magma is sucked up under volcanoes.

Further scrutiny of Fig. 1.1, the photo-elastic assembly of discs, shows another interesting feature: the force distribution is very heterogeneous. Some regions are entirely force-free, while other regions experience high inter-particle forces that frequently — but not exclusively — line up to form 'force bridges'. The variability in contact forces points to an accompanying variability in local deformations. Here is something that will prove very important in the study of the mechanics of granular media that are not packed in a regular lattice (which is only possible if there is only one grain size or for a very particular combination of sizes), which is the norm in any naturally occurring sample: *granular media are intrinsically heterogeneous*. The consequences of this for the mechanics of a granular assembly will be explored in forthcoming chapters.

When the material reaches the plateau of the stress ratio in Fig. 1.2 another feature may become apparent. As the tangent modulus becomes poorly defined the material may find, depending on the precise boundary conditions, a mode of motion that is localised. Such 'rupture layers' and 'failure' are very important for the engineering community, as illustrated in the example of a landslide occurring as described earlier in this section.

Literature on soil mechanics is plentiful: [Lambe and Whitman, 1969] is a classic text, as is [Terzaghi, Peck and Mesri, 1996]; [Powrie, 2004] is a more modern textbook.

1.2 The isostatic state and jamming

A static packed assembly of grains in contact confined by a compressive stress is equivalent to a network of forces. As it is static, force and moment equilibrium will hold. The question being addressed in this section is: how many forces in the network can be specified in such a way that force and moment equilibrium *alone* are sufficient to determine them, given the detailed geometry of the conformation?

A few conditions need to be laid down to come to a non-trivial answer. The first is that a regular packing is excluded from the analysis; an assembly in a regular packing satisfies certain symmetry rules which need to be imposed in addition to the equilibrium equations. Thus, a medium that consists of identical spherical particles is not accounted for at this stage. Rather, a polydisperse grain-size distribution is envisaged, making for a random packing. Alternatively, rough particles may make up the assembly. No isotropic condition imposition is necessary, though this is often (sometimes tacitly) assumed in the literature. Furthermore, it is assumed that the assembly is very large, so that the number of forces on the perimeter of the sample is small compared to the number of forces in the network. Basically, any condition that somehow constrains the forces in the network is excluded for the moment, implying that the equilibrium equations alone are sufficient to do the analysis. Specific constraining assumptions are discussed below.

In a random packing with N interacting particles in d dimensions there are Nd force equilibrium equations, as each particle that participates in the network is in equilibrium. The force moment equilibrium for each particle provides $d(d-1)/2$ equations, so for N particles there are $Nd+Nd(d-1)/2=Nd(d+1)/2$ equations. Each contact force will have d components and is shared by two particles. Equating the two gives the result that it is possible to calculate $N(d+1)$ forces, or an assembly coordinate number, that is the number of contacts per particle, of $N_{c,iso}=d+1$ forces on average (the subscript 'iso' refers to the isostatic case). Note that this average pertains to particles that participate in the force network. It is well possible that a fair percentage of particles have no contact and these obviously do not contribute to the evaluation of the isostatic coordinate number.

When there are more force-carrying contacts, the equilibrium equations alone will not be able to permit the calculation of the forces. The system is then *statically indeterminate*. When there are fewer than $d+1$ contact forces per particle there are more equations than unknowns and the system cannot be in static equilibrium. The isostatic state is therefore a very precarious, marginally stable state. The slightest disruption that results in the loss of even one contact will make the structure change until the number of force-carrying contacts is at least equal to the required number.

The number $d+1$, which equals 3 in two dimensions and 4 in three dimensions, compared to any experimental result for a densely packed material shows that for practical purposes the statically indeterminate state is much more relevant. However, the analysis changes somewhat when constraints are imposed. So, the assumption of randomness is still maintained, but a constraint may follow from the fact that certain contacts slip. In that case the nature of friction must be considered.

Particles interact via their surfaces and these need not be smooth. As long as the surfaces are 'infinitely sticky' the force component that is tangential to the surface is free to take any value. In cases where slip is relevant, a Coulomb-type constraint reigns in which the magnitude of the tangential force remains equal to μ_s times the normal force. Contact forces must then be classified according to those that stick and those that slip. Let the ratio of slipping contacts in the assembly be given as a fraction f_μ of all the contact forces, then the number of sticking forces populates a fraction $1-f_\mu$. The number of equations and unknowns now stack up as follows:

Nd force equilibrium equations

$Nd(d-1)/2$ moment equilibrium equations

$f_\mu NN_{c,iso}/2$ slipping conditions

$NdN_{c,iso}/2$ unknown contact force components

Equating the number of equations with the number of unknowns gives the result that the coordinate number per particle is

$$N_{c,iso} = \frac{d(d+1)}{d-f_\mu} \tag{1.1}$$

The implication is that as the fraction of slipping contacts increases, the number of contacts that need to be accommodated in the assembly will go up. When all contacts slip ($f_\mu = 1$) in both two and three dimensions the value of $N_{c,iso}$ is 6.

A very special case occurs when there is no friction and the particles are perfectly spherical or discs. In that case all forces are normal to the contact surfaces and the moment equations become redundant: $N_{c,iso} = 2d$.

Again it is emphasised that these considerations only pertain to the particles that participate in the force network.

An experiment may be envisaged in which the particle assembly starts of as very dilute; it is then compacted (say, isotropically). There comes a point in this process when the particles begin to touch. When the number of particles that touch is sufficient for the medium to be on the edge of static equilibrium the assembly is said to 'jam'. Compressing the assembly further will involve the compression of enduring contacts and therefore the development of a stress. The packing density at which the jamming transition takes place may be determined in numerical simulations. The outcome depends on assumptions on polydispersity (for spheres and discs), the details of the simulation method (gravity on or off, for example) and — indeed — the precise definition of the jamming density. Therefore, the concept of a 'jamming transition density' may only have approximate meaning.

Moreover, the analysis above shows that the number of contacts that can be supported in the isostatic state depends strongly on the fraction of the contacts that slip. In numerical simulations parameters can be tightly controlled to set the value of inter-particle friction (infinite and zero are popular choices), as well as the shape of the particles that participate in the simulation and the strain path that is employed. In any physical experiment with natural or manufactured particles, however, these parameters are not so easily controlled. The inter-particle friction coefficient, for example, may exhibit natural variation and therefore take a range of values; furthermore, particles tend to be rough and only approximately spherical.

A further question is whether an assembly of particles can be 'partly isostatic', that is that regions within the assembly can be distinguished for which the numbers of equilibrium equations equals the number of forces while there are also regions where there are fewer. Doubtlessly conditions can be found, involving factors such as closeness to the jamming condition and nature of the particle interaction (for example rough or smooth), where this is the case. In the references the relevant literature is highlighted. One aspect that comes to the fore in these papers is the need to distinguish *fluctuations* in the local geometry. For dense assemblies, where the intention is to obtain a stress-strain relation, the most convenient approach is to introduce an inter-particle interaction and to develop the theory further taking account of the fluctuations in that context.

An interesting feature of the present discussion is an historical perspective. The conditions for isostaticity were originally laid out by [Maxwell, 1864]. In fact, Maxwell's text employs identical arguments as the one at the beginning of this section. A fully elaborated theory of static indeterminacy was produced by Mohr in 1874, see [Mohr, 1906]. Not until a century later did these concepts find their way into the literature of granular mechanics. In the early 2000s a more rounded view of the subject became available and the notion that sliding friction may influence the theory. A great help has been the development of simulation methods so that the jamming transition may be studied 'experimentally'. Jamming under non-isotropic conditions has been included more recently.

An extensive overview of the jamming transition is described by [Liu and Nagel, 2010]. Stresses in an isostatic assembly are derived by [Blumenfeld, 2007] and in this paper some other problems regarding the concept of isostaticity are also highlighted. Non-isotropic compression and jamming (with physical experiments) is discussed by [Bi *et al.*, 2011]. An exhaustive list of publications relevant to this subject is somewhat outside the scope of this text, however most relevant ones are in the references mentioned.

1.3 The statically indeterminate case and computer simulations

The next problem must be how the contact forces are going to be solved in the statically indeterminate state. In this case there are more force variables than force and moment balance equations (and more contacts per particle than $N_{c,iso}$). A solution is possible when a constitutive equation is introduced. Such an equation gives the relation between force and displacement difference between particles (particles may also rotate and this too needs to be incorporated in the constitutive equations). It necessarily implies that the particles are deformable. This may be counterintuitive as particles are frequently thought of as rigid (sand grains, for example, would appear to be very stiff). More precisely, a rigid limit can be thought of when the stiffness of the particles is very much greater than the pressure associated with the stress in the assembly. However, allowing for small indentations during particle contact resolves the issue of static indeterminacy. Here is a list of unknowns and equations for all the particles that participate in the force network.

Unknowns

Nd particle displacements
$Nd(d-1)/2$ particle rotations
$NdN_c/2$ contact forces

Equations

Nd force equilibrium equations
$Nd(d-1)/2$ moment equilibrium equations
$NdN_c/2$ contact force — relative particle displacement and rotation relations (the contact laws)

The number of unknowns (that is, the displacements and rotations) is equal to the number of equations and (assuming no mathematical pathologies) a solution may be constructed. The reader may now be surprised that there is no mention of a torque constitutive equation. There is an underlying assumption here (which is similar to the rigidity assumption) that the contacts may be thought of as *point contacts*. A point contact cannot transmit a torque. So, unless the particles are very deformable — and the contact area may acquire an appreciable value — this aspect may be neglected. A problem would arise when the grains in the force network are so irregularly shaped that two neighbouring particles may share more than one contact. In that case, of course, a torque may be transmitted. In principle the theory can be easily amended to account for a complication like that by introducing a particle contact(s) torque in addition to the contact forces and an extra set of constitutive equations relating particle rotation to the transmitted torque. This is not followed up here, where it is assumed that the particles are hard (though slightly deformable) and share at most one contact.

The set of equations, as outlined above, can be solved using *computer simulations* and in that way displacements and rotations of the particles in an assembly may be determined under suitably chosen boundary conditions.

In the literature it is only very rarely that a procedure is encountered in which a *quasi-static solution* (QS) is constructed. Nonetheless, it *is* possible to do this. [Koenders and Stefanovska, 1993] show an approximation method, based on a least-squares approach of the force and moment equilibrium equations for an elasto-frictional material in two dimensions. The result for a biaxial cell test are very similar to the ones measured by, for example, [Konishi, 1978]. The latter is an experiment on photo-elastic discs – see Fig. 1.1. The statistics of the micro-mechanical variables are faithfully reproduced. These include the mean contact distribution and the distribution of the slipping contacts as the test progresses. Macroscopic features, such as the stress ratio reaching a maximum and the occurrence of dilatancy are also found.

Despite the relative success of this method, it has not been pursued by many other researchers, who have preferred *dynamic methods*.

These are obviously attractive if, in addition to slow changes to an assembly in the high contact number régime, faster changes and granular flow also need to be studied. To accommodate the dynamics, a particle mass and moment of inertia terms need to be introduced to the equilibrium equations, so that a full Newtonian set of equations is processed. To solve Newton's equations simultaneously with evolving contact properties, such as detecting new contacts and deleting old ones, for all particles in a large assembly (say, $N > 1000$) requires a massive computer effort. In a molecular dynamics method, called the *Discrete Element Method* (*DEM*), a sequential approach is taken, using a small time step and moving and rotating the particles in the assembly one at a time and after that updating the contact properties. If the time step is small enough, this would be equivalent to a simultaneous solution. The method was first introduced by [Cundall and Strack, 1979] — a two-dimensional version of the DEM. Since its inception it has been developed further and has been expanded to three dimensions, more complicated contact laws and extensions to include more general boundary conditions, including periodic ones. More complicated particle shapes with rough boundaries have been included in an attempt to model realistic, natural conditions. The method has had a tremendous influence on the development of the subject, not least because proposed theoretical models in which micro-mechanical assertions are put forward could be tested against computer simulations.

Free software and many informative documents are available, so researchers can run their own simulations [Yade, 2019].

It is fair to say that reporting on the results of the method has not always been entirely complete. It is also the case that in some instances the reporters have been arrogant in asserting that the simulation results are superior to physical experiments, though it *is* true that in the computer certain boundary conditions can be simulated that are very difficult to realise with a laboratory apparatus, see for example [Thornton, 2000]. Consistent examples of papers on simulations that use the method (and discuss some of the difficulties with it) are by [Thornton and Antony, 1998] and a very informative paper by [Thornton and Sun, 1993]. Further useful papers, showing the potential and increased subtlety of the method, are by [Ferellec and McDowell, 2010], [Macaro and Utili, 2012] and [McDowell and Li, 2016]. This little list is illustrative only and does by no means justice to the extent to which papers on this subject have been published. There must be many thousands.

A computer method that lies somewhere between QS and the DEM is the *Contact Dynamics* (CD) method. The background to this is the following. The time step in a fully dynamic implementation of the equations of motion needs to be so small that it is adequate to follow the changes in the contact laws. The latter allow for a small indentation in what are essentially rigid particles. The problem with the contact laws is that they are highly non-linear and therefore a large number of time steps is required to model, say, a collision between two particles. In the CD method the accelerations are not calculated, but the particles are subject to a velocity field. The latter can change abruptly, both in direction and magnitude. This is so-called non-smooth motion. For the method to work, the motion during the encounter between two particles is integrated, taking into account the non-linear contact laws. The inputs to any collision encounter are the velocities of the two participating particles, while the output consists of the velocities after the encounter has taken place. The actual integration cannot be done exactly, but certain estimates have to be made. These have been elevated to a high art by the CD community and it is generally assumed that the method is no less accurate than the DEM method. More specifically, the propagating error introduced by the exceedingly small time step in a fully dynamic program may well be of the same order of magnitude as the error incurred in the approximations in the integration method in CD. *Any* computational method is approximate in some sense. However, CD is much faster than DEM, as rather larger time steps can responsibly be taken. Relevant references for this method are by the inventors of the method [Jean and Moreau, 1992] and [Jean, 1999], as well as an informative introductory paper by [Radjai, 2008]. Again, as with the treatment of the DEM before, there are many more papers that could be quoted, especially as the method has gained in popularity in recent times.

1.4 Contact laws

When drawing up a suitable constitutive law for contact relating contact displacement and contact force the first thought should be 'what is it meant to achieve?' In molecular studies and studies of small particles in liquids very sophisticated interactive relations have been put forward that account for surface potential effects and quantum mechanical interactions. These relations are highly non-linear and allow for both repelling and attractive phenomena. However, in dealing with larger particles a simple law that just ensures that the particles only overlap by a very small amount would appear to be sufficient. The difficulty with increasing sophistication is that it requires more and more parameters, which may be difficult to measure. Also, the benefit of more complex laws is marginal. The need for a contact law arises from the existence of a statically indeterminate state. The first goal is to fix this problem by simple means and get some insight in the properties of such systems. Added complexity can be inserted later as a refinement.

Any two surfaces that touch one another could in a first approximation be assumed to repel one another as springs. This gives a relation for the normal force between the surfaces that is characterised by merely the spring constant k. The latter will generally be a function of the contact force itself. The non-linearity that is associated with that gives rise to the need to introduce incremental contact laws (the need for incremental laws will be discussed in more detail below). So, if the normal displacement $D_\perp$ is related to the normal force $F_\perp$ via a spring constant

$$F_\perp = k(F_\perp)D_\perp$$

Then the incremental law reads

$$f_\perp = \frac{\partial k(F_\perp)}{\partial F_\perp} d_\perp$$

The function $k(F_\perp)$ may contain a number of features. In addition to the non-linearity the incremental spring constant may be either assumed to be entirely elastic or reflect certain plastic effects (that is, have different values for loading and unloading).

1.5 The frictional interaction

One effect that is without doubt very important in the constitutive contact law is the effect of *friction* and to introduce that the normal force alone is insufficient; a tangential force-displacement rule must be added to the description.

The friction effect is obviously plastic. When the force ratio (that is the magnitude of the tangential force to the normal force) reaches a certain value μ_s, persistent further motion will not change it; a constraint has become active that keeps the force ratio constant. This was established by [Coulomb, 1785] (based on measurements by [Amontons, 1699], see [Heyman, 1972] for the history of the subject and many more references). The concern here is essentially with dry friction. [Bowden and Tabor, 1956] treat the subject from an engineering standpoint and also extend their treatment to include effects of lubrication. On unloading the contact may recover its elastic properties, though not necessarily with the same elastic constant as the loading curve. The process is illustrated in Fig. 1.3, where $F_{\|} / F_{\perp}$ is shown as a function of the tangential contact displacement $D_{\|}$. In this figure the spring constants for loading and unloading are taken as constants; when a nonlinearity is taken into account the straight loading and unloading lines become curved.

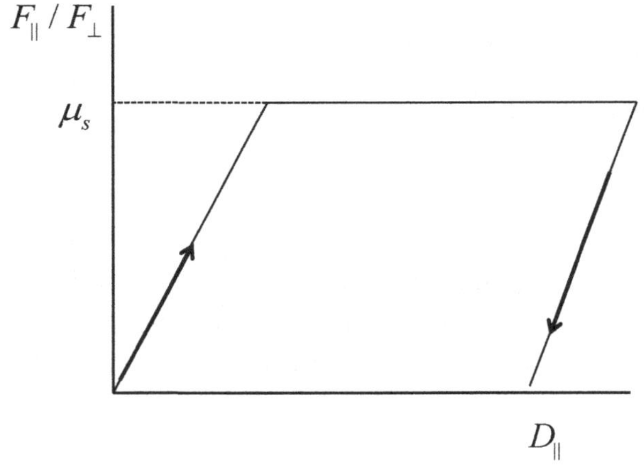

Figure 1.3. Illustration of the Coulomb friction principle.

It is clear that when behaviour like this is encountered an incremental formulation is necessary. The normal and tangential motion become coupled, so a general form for the incremental contact response relates the force increment to a displacement increment

$$\begin{pmatrix} f_\perp \\ f_\parallel \end{pmatrix} = \begin{pmatrix} k_{\perp\perp} & k_{\perp\parallel} \\ k_{\parallel\perp} & k_{\parallel\parallel} \end{pmatrix} \begin{pmatrix} d_\perp \\ d_\parallel \end{pmatrix}$$

The elements of the matrix are the spring constants. Some properties of these are easily determined.

In the *elastic state* there is an incremental potential, $\frac{1}{2}\left(k_{\perp\perp}d_\perp^2 + \frac{1}{2}\left(k_{\perp\parallel} + k_{\parallel\perp}\right)d_\perp d_\parallel + k_{\parallel\parallel}d_\parallel^2\right)$, from which the force increment is obtained by partial differentiation. It follows immediately that for this case the matrix must be symmetric: $k_{\perp\parallel} = k_{\parallel\perp}$. If, in addition it is assumed that normal and tangential increments are uncoupled, it follows that the off-diagonal elements vanish. This does not preclude anisotropy with respect to the direction of the contact surface, so $k_{\perp\perp}$ need not be equal to $k_{\parallel\parallel}$. If they are assumed to be equal then the one-parameter model $\mathbf{k} = k_{\perp\perp}\boldsymbol{\delta}$ follows, which has the pleasant property that it is invariant under rotation, so the direction of the contact is irrelevant (the Kronecker delta $\boldsymbol{\delta}$ is formally introduced in Section 2.2). For contacts that only interact through the normal force (frictionless) $k_{//} = 0$.

In the frictional sliding state an additional force increment added to the state $\left(F_\perp, F_{//}\right)$ should leave the ratio $F_{//} / F_\perp$ constant at the value of μ_s. Taking $F_\perp$ and $F_\parallel$ both positive, leads to the following

$$\mu_s = \left|\frac{F_\parallel + f_\parallel}{F_\perp + f_\perp}\right| \cong \left|\frac{F_\parallel}{F_\perp}\left(1 + \frac{f_\parallel}{F_\parallel}\right)\left(1 - \frac{f_\perp}{F_\perp}\right)\right| = \left|\frac{F_\parallel}{F_\perp}\right|\left(1 + \frac{f_\parallel}{F_\parallel}\right)\left(1 - \frac{f_\perp}{F_\perp}\right) \cong \mu_s\left(1 + \frac{f_\parallel}{F_\parallel} - \frac{f_\perp}{F_\perp}\right)$$

In other words

$$\frac{f_\parallel}{F_\parallel} - \frac{f_\perp}{F_\perp} = 0 \rightarrow f_\parallel - \mu_s f_\perp = 0$$

This constrains the elements of the matrix $\mathbf{k}$ by the additional relation

$$k_{\parallel\perp}d_\perp + k_{\parallel\parallel}d_\parallel - \mu\left(k_{\perp\perp}d_\perp + k_{\perp\parallel}d_\parallel\right) = 0$$

which must hold for arbitrary displacements, hence

$$k_{\parallel\perp} - \mu_s k_{\perp\perp} = 0; \; k_{\parallel\parallel} - \mu_s k_{\perp\parallel} = 0$$

So, for this case the matrix **k** takes the form

$$\mathbf{k} \rightarrow \begin{pmatrix} k_{\perp\perp} & k_{\|\|} / \mu_s \\ \mu_s k_{\perp\perp} & k_{\|\|} \end{pmatrix}$$

In many instances the increase in the tangential force increment for purely tangential motion is negligible, implying that $k_{\|\|} = 0$. The frictional state is then entirely described by two parameters, $k_{\perp\perp}$ and μ_s.

When $F_{\|}$ is negative, μ_s is replaced by $-\mu_s$; otherwise the relations remain the same.

Unloading from the frictional state is detected by checking what the response would have been for an elastic increment (this could in principle be brought about by an increase in the normal force). If this decreases the magnitude of the tangential to normal force ratio, the next increment should be evaluated using the (unloading) elastic law. Therefore, *the frictional interaction is predictive, but must always be followed by a verification.*

Friction in two dimensions is covered in the literature. [Ruina, 1980] and [Ruina, 1983] discusses the sliding state once the initial friction criterion is passed. On continued motion the value of μ_s falls by a small amount — the friction is said to change from a static value to a kinetic value. In addition, an extra stress that is proportional to the speed of continued tangential motion needs to be introduced (this effect is sometimes known as the Ruina–Dieterich law: [Dieterich, 1979, 1981]). It should be emphasised that the measurements that underlie this law are done on blocks of rock material. In these experiments there are always many contacts at the same time, while for the present application two particles share one contact, which is approximately a point-contact, that is, a very small contact area between two convex surfaces. Direct application of the Ruina-Dieterich law may therefore not be appropriate.

While the frictional effect has been measured extensively, the actual mechanism of the contact mechanics that lead to friction is relatively unexplored. [Villagio, 1979] has put forward some interesting ideas, though they have so far not been widely followed up.

1.5.1 Friction in three dimensions

The exposition given above is idealised in that the motion and force parameters all operate in a plane. To some extent that is a view justified by the fact that the frictional interaction takes place on the surface of two bodies in contact. The unit normal of the surface is $\mathbf{n}$ and if the force across the surface is $\mathbf{F}$, the normal component is the inner product $F_\perp = \mathbf{F}\bullet\mathbf{n}$. The tangential force is then $\mathbf{F}_\parallel = \mathbf{F} - (\mathbf{F}\bullet\mathbf{n})\mathbf{n}$. The sliding friction criterion may now be expressed as $\sqrt{\mathbf{F}_\parallel \bullet \mathbf{F}_\parallel} = \mu_s F_\perp$. This relation represents a cone, as illustrated in Fig. 1.4.

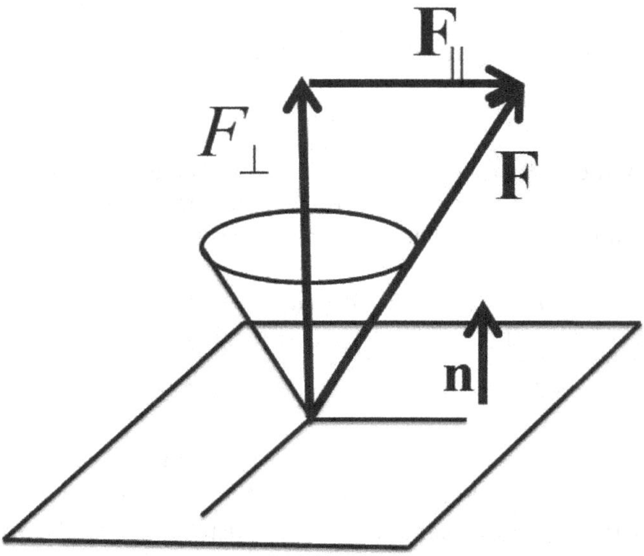

Figure 1.4. Friction cone. The opening angle is $2\tan^{-1}\mu_s$.

The procedure for obtaining the incremental interaction in the sliding state is the same as before. Basically, the force vector must be constrained to move on the surface of the cone.

The most convenient way of making progress is now to choose a coordinate frame that is aligned with the forces. One unit vector — $\mathbf{n}$ — is already in place; of the other two one is chosen to be aligned with

$\mathbf{F}_{||}$ and the other one normal to that (as well as normal to $\mathbf{n}$). The former is called $\mathbf{n}_{||}$ and the latter $\mathbf{n}_{\Diamond}$ (this vector is sometimes called the *binormal*). In this frame the force and the force increment have components

$$\mathbf{F} \to \begin{pmatrix} F_{\perp} \\ F_{||} \\ 0 \end{pmatrix} \Rightarrow \mathbf{f} \to \begin{pmatrix} f_{\perp} \\ f_{||} \\ f_{\Diamond} \end{pmatrix}$$

The sliding friction criterion becomes

$$\sqrt{(\mathbf{F}_{||} + \mathbf{f}_{||}) \bullet (\mathbf{F}_{||} + \mathbf{f}_{||})} = \mu_s (F_{\perp} + f_{\perp})$$

Expanding up to first order in the increments gives

$$f_{||} = \mu_s f_{\perp}$$

This is exactly the same relation as for the two-dimensional case and the implications for the incremental force-displacement relation are also obtained in a similar fashion. The resulting interaction is

$$\begin{pmatrix} f_{\perp} \\ f_{||} \\ f_{\Diamond} \end{pmatrix} = \begin{pmatrix} k_{\perp\perp} & 0 & \mu_s^{-1} k_{||\Diamond} \\ \mu_s k_{\perp\perp} & 0 & k_{||\Diamond} \\ k_{\Diamond\perp} & k_{\Diamond||} & k_{\Diamond\Diamond} \end{pmatrix} \begin{pmatrix} d_{\perp} \\ d_{||} \\ d_{\Diamond} \end{pmatrix}$$

The question now is whether there is a coupling between the third direction and other two directions. If the third direction operates entirely independently then all coefficients with a $\Diamond$ vanish other than the diagonal term $k_{\Diamond\Diamond}$, which is probably some fraction of the normal diagonal coefficient $k_{\perp\perp}$. If, however, the tangential force influences the behaviour in the third direction then a number of extra parameters need to be taken into account. For point contacts these parameters are very difficult to measure.

1.6 Contact laws in terms of material parameters

A question that is particularly of interest to the simulation community concerns the matter whether the spring constant can be related to the material properties of the particles. The prime candidate for such a theory is the Hertzian contact theory, which deals with two elastic bodies that are being compressed together — [Hertz, 1882]. The details are in [Landau and Lifschitz, 1976] and more extensively in the [Johnson, 1985] book on contact mechanics.

For two spheres pressed together by a force $F_\perp$ the distance between the centres of two spheres with radii R and R' is reduced by an amount $D_\perp$

$$D_\perp = F_\perp^{2/3}\left[Q^2\left(\frac{1}{R}+\frac{1}{R'}\right)\right]^{1/3},$$

where the parameter Q contains the elastic constants (Young's moduli E, E' and Poisson's ratios ν, ν') of the materials of which the solid bodies are made:

$$Q = \tfrac{3}{4}\left(\frac{1-\nu^2}{E}+\frac{1-\nu'^2}{E'}\right)$$

An obvious aspect of this force-indentation formula is that the force-displacement relationship is non-linear. An incremental relationship is easily obtained

$$f_\perp = \tfrac{3}{2}\frac{F_\perp^{1/3}}{Q^{2/3}}\left[\left(\frac{1}{R}+\frac{1}{R'}\right)\right]^{-1/3} d_\perp \equiv k_{\perp\perp}(F_\perp)d_\perp$$

A relationship between the proposed spring constant and the material parameters of the particles could then be proposed as some assembly average value of $k(F)$. For simulation purposes that would be very unsatisfactory and the vast majority of simulists code for the original relationship between $D_\perp$ and $F_\perp$. For analytical modelling, however, where interactive properties frequently appear as sums over nearby particles, an averaging approach may be convenient for theoretical purposes.

In two dimensions, relating a spring constant to material properties can be ascertained by looking at the compression of two cylinders along their axes. The Hertzian relationship is, see [Puttock and Thwaite, 1969] and [Williams and Dwyer-Joyce, 2001]

$$D_\perp = \frac{4}{3\pi}\frac{F_\perp}{\ell}Q\left\{1+\ln\left[\frac{6\pi\ell^3}{F_\perp Q}\left(\frac{1}{R}+\frac{1}{R'}\right)\right]\right\},$$

where ℓ is the length of the axes of the cylinders and R, R' are the cylinder radii. Note that the expression essentially depends on the force per unit length.

This expression is also non-linear, but not easily employed, because inverting it (to give $F_\perp$ as a function of $D_\perp$) leads to extra numerical work.

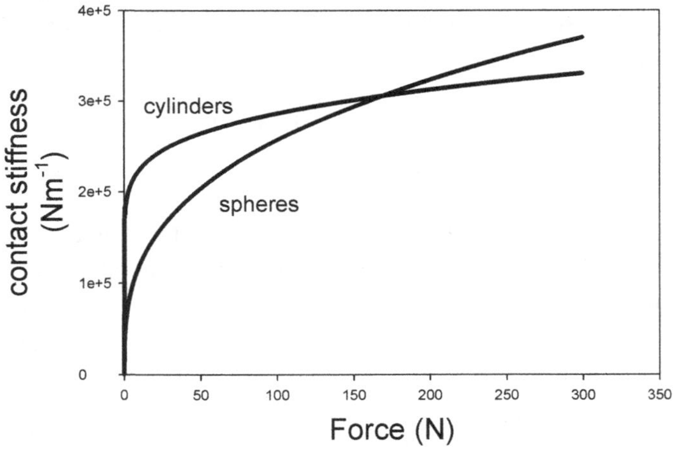

Figure 1.5. Contact stiffness as a function of the applied compressive force for two spheres and two cylinders. The parameters used are as follows: $Q = 10^{-8}\,m^2N^{-1}$; all radii $.01m$ and the cylinder length $\ell = .01m$.

The expression can be differentiated with respect to $D_\perp$ and then $\partial F_\perp / \partial D_\perp$ as a function of $F_\perp$ may be obtained,

$$\frac{\partial F_\perp}{\partial D_\perp} = k_{\perp\perp} = \frac{3\pi\ell}{4Q\ln\left[\dfrac{6\pi\ell^3(R+R')}{QRR'F_\perp}\right]}$$

A comparison of the contact stiffnesses for the cases of two spheres and two cylinders is illustrated in Fig. 1.5. It is seen that there is a very substantial variation in the result, especially lightly loaded contacts have a vanishingly small incremental stiffness.

The assumption made here is that the contact areas are immaculately clean and that the contact is perfectly smooth. For two particles made of a natural material — sand particles, for example — that assumption is obviously severely contestable. Further research on two fractal surfaces pressed together is reported in [Hanaor *et al.*, 2015]. As the only purpose of the contact stiffness is to hinder two particles from overlapping (and do so in a controlled manner), it could be argued that *any* stiffness is fine, as long as the indentation is such that only a very small overlap (compared to the typical radius of the particles) is effected at the typical contact force regime in the assembly. It is also noted that for particles composed of natural materials a number of plastic effects can be expected (including breakage). Therefore, frequently, researchers just take a constant value for the contact stiffness and add friction, for example [Kuhn, 1999]. This is computationally simple and achieves the purpose of rectifying the problems of determining the contact forces in the case that the assembly is not in an isostatic state, at the expense of some physical realism. This is a perfectly reasonable thing to do.

In some sense the details of the normal interaction are not that critical. The tangential stiffness, including frictional effects can be added to the interaction. There are various approaches. A well-known one takes account of slip in an annulus inside the contact area. The extent of the annulus depends on the applied force ratio. A fair amount of ink has been spilt over this problem; quoting [Johnson, 1985]: 'In a paper of considerable complexity, [Mindlin, 1949], [Mindlin and Deresiewicz, 1953], have investigated the changes in surface traction and compliance between spherical bodies in contact arising from the various possible combinations of incremental change in loads: normal force increasing, tangential force increasing; normal force decreasing, tangential force increasing; normal force increasing, tangential force decreasing; *etc.*' The parameters needed are the material stiffness, Poisson's ratio and a friction coefficient. Despite the complexity of the calculations the result in terms of incremental contact law is not dissimilar from the one obtained from the phenomenological approach as pursued above. All the notes regarding the idealisation of the problem, and therefore the question marks that accompany an application to the 'dirty' materials of which the real world is composed, are relevant again. It could also be argued that an analysis meant to explain friction based on the assumption of a friction coefficient is tautological, at best adding details to the mechanism.

1.7 Interaction for small particles in a fluid environment

This section deals with small particles, micron- and sub-micron-sized, in a fluid environment. The question is how such particles interact when they come close together. Applications in chemical and environmental engineering (especially filtration), cosmetics, the mechanics of clay, etc. are envisaged. In these applications dense cakes of small particles are created and subsequently manipulated by either sedimentation or filtration methods.

The particles are solids, implying that the constituent molecules are in some sort of crystal structure. On the boundary of the particle solid the crystal structure meets the fluid; the crystal arrangement suddenly ends. There is then a discontinuity in the electric charge distribution, which is accommodated by the recruitment of the ions in the fluid near the boundary into a compensating configuration. The fluid molecules, however have a far greater mobility than those in the solid. Moreover, their equilibrium state — far from the solid boundary — is determined by the type of molecule in the fluid and its temperature.

The mobilisation of the ions in the fluid is achieved by either turning the dipoles of the fluid molecules in the direction of the solid boundary, or by attracting or repelling ionic charges. This can only be partially successful, as the thermal motion tends to make the alignment less effective. Also, if in a fluid a layer of molecules has a more or less aligned dipole moment, the next layer of fluid will respond by turning its dipoles in the opposite direction in order to achieve charge neutrality. Thus, a *double layer* is created. The electrical potential in the fluid as a function of the distance from the boundary will be a declining function.

Now, if two particles are brought together there are two declining potentials and the charges inside the fluid will act on that, effectively causing a repulsive interaction. This is called the *double layer interaction* and it is part of a multi-aspected interaction, the so-called DLVO theory — named after its main contributors Debije, Landau, Verwey and Overbeek. The analysis of the complete theory involves a large number of approximations, basically taking account of the repulsive double-layer interaction and an attractive van der Waals interaction.

The literature on this subject is vast. The classic is [Kruyt and Overbeek, 1969]. Good textbooks that treat the basics and a plethora of applications are [Hunter, 1987, 2001].

The theory of the double layer interaction is extremely well-researched in the colloid literature and all that needs to be done here is to communicate the results.

A measure for the thickness of the double layer is some chosen multiple of κ^{-1} and κ is approximately

$$\kappa = \sqrt{\frac{e^2 n^{(0)} Z^2}{\varepsilon k_B T}},$$

where e is the electron charge, $n^{(0)}$ the bulk concentration of ions, Z the valency of the ions, ε the electrical permittivity of the fluid, k_B Boltzmann's constant and T the absolute temperature. If there are more than one type of ions in the fluid the concentration and valences are simply summed. Now, the interaction between two particles depends on the separation of the particles H and the parameter κ; the simplest non-dimensional combination is $H\kappa$. Thus, the double layer interaction is a function of $H\kappa$. The actual form of the interaction is exposed in two approximations involving the particle radius a. The first approximation pertains to the case in which κa is large (say, larger than 10). Defining the surface potential as ψ_0, the interactive potential is

$$V(H) = 2\pi\varepsilon a\psi_0^2 \ln\left(1 + e^{-\kappa H}\right)$$

The second approximation is relevant for $\kappa a < 5$, in which case the interactive potential takes the form

$$V(H) = \frac{4\pi\varepsilon a^2 \psi_0^2}{H} e^{-\kappa H}$$

In these formulas the surface potential as ψ_0 depends on the type of surface and the ionic content of the fluid. The interactive force is obtained from $-\partial V / \partial H$.

The van der Waals contribution has also been evaluated. Here the interactive potential for two equal particles is given with A_{12} a constant called the Hamaker constant (the analysis is due to [Hamaker, 1937])

$$V(H) = -A_{12} \frac{\left(\frac{H}{a}\left(\frac{H}{a}+2\right)^2 \ln\left(\frac{2\frac{H}{a}}{\frac{H}{a}+2} \right) + 3\frac{H}{a} + 2 \right)}{6\frac{H}{a}\left(\frac{H}{a}+2\right)^2}$$

If the two particles are very close together ($H / a \ll 1$) then this reduces to

$$V(H) = -\tfrac{1}{6} A_{12} \ln\left(\frac{H}{a}\right) - \tfrac{1}{12} A_{12} \frac{a}{H}$$

The contributions from the double layer interaction and the van der Waals interaction can be added to give the main contributors to the DLVO theory. The total effect depends on the coefficients, which reflect the exact type of system that is relevant. An example of the sum of the two contributions is given in Fig. 1.6.

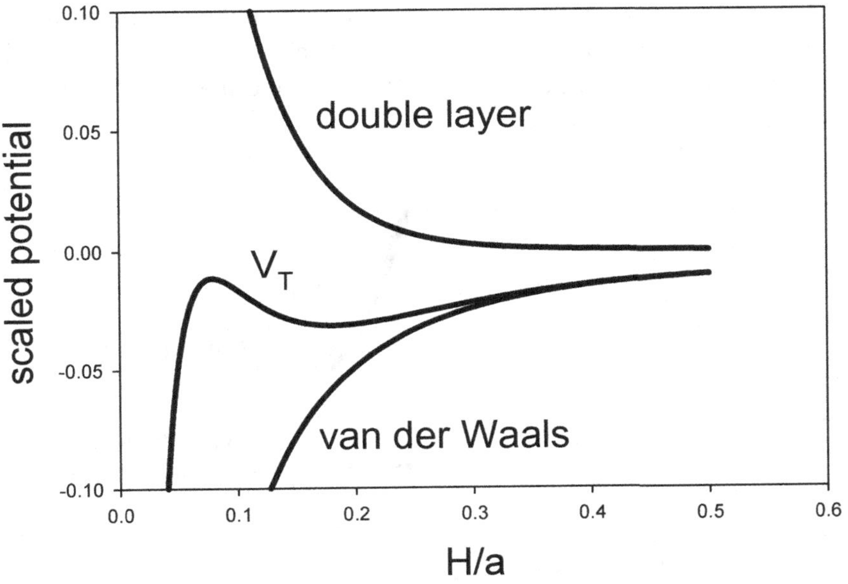

Figure 1.6. Illustration of the potential V_T scaled to $2\pi\varepsilon a\psi_0^2$ for a value of $\kappa a = 20$ and $A_{12} / \left(2\pi\varepsilon a\psi_0^2\right) = 0.3$.

The example in this graph is chosen to highlight some features. Some numbers are relevant. Suppose the particle radius is $0.2\,\mu m$, then the double layer thickness is $\kappa^{-1} = 10nm$. For distances less than a few nanometres the theory is unreliable. In the figure that corresponds to $H / a \approx 0.03$. The sum of the two contributory potentials V_T is then not accurately represented for very small H / a. Keeping that in mind, two features of the combined potential are clearly visible. Firstly, there are two attractive wells, one very close to the particle (where the theory is not valid) and one around $H / a = 0.18$. Secondly, moving the particles closer together from the latter minimum, there is a potential to overcome. It must be pointed out that these features are specific to the choice of parameters that has been made.

For much thicker double layers there are no potential minima in the relevant range and the interactive force is always repulsive. This is illustrated in Fig. 1.7 where $\kappa a = 3$. Note that the interaction is highly non-linear.

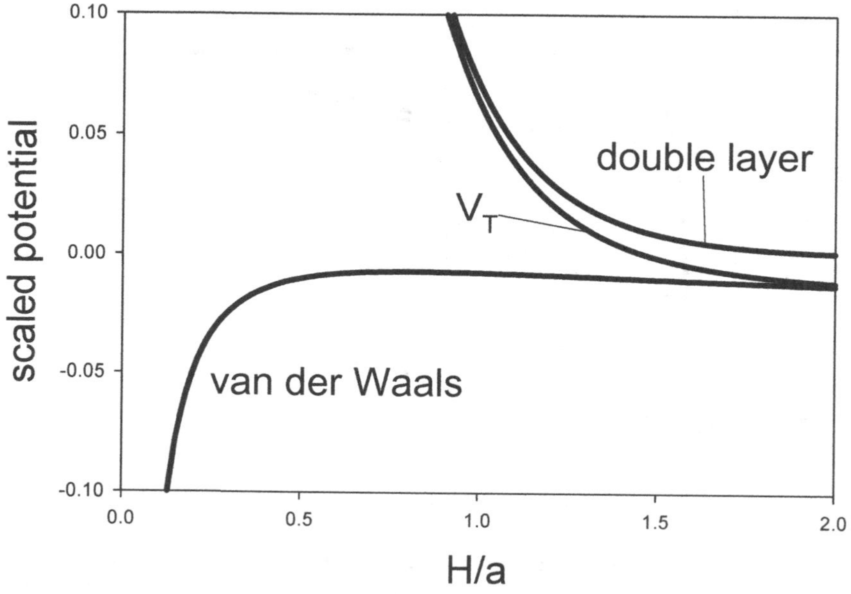

Figure 1.7. Illustration of the potential V_T scaled to $2\pi\varepsilon a\psi_0^2$ for a value of $\kappa a = 3$ and $A_{12} / \left(2\pi\varepsilon a\psi_0^2\right) = 0.3$.

The plethora of behaviours of colloidal substances is largely due to the variety of possible outcomes for the interactive potential curve and whether there are minima or maxima in the ambient mechanical (and thermal) environment.

One consequence of the existence of an interactive potential is that there is always a force active between neighbouring particles and as a result considerations relating to the isostatic state are not as acute as in the case of an interaction that is solely due to contact.

References

Amontons, G. (1699) *De la resistance causée dans les Machines, tant par les frottemens des parties qui les composent, que par roideur des cordes qu'on y employe, & la maniere de calculer l'un & l'autre* (On the resistance caused in machines, both by the rubbing of the parts that compose them and by the stiffness of the cords that one uses in them, and the way of calculating both), *Mémoires de l'Académie royale des sciences*, in: Histoire de l'Académie royale des sciences, pp. 206–222.

Bi, D., Zhang, J., Bulbul Chakraborty, B. and Behringer, R. P. (2011) Jamming by shear. *Nature* **480** 355–358.

Bowden, F.P. and Tabor, D. (1956) *Friction and Lubrication.* London: Methuen & Co Ltd.

Blumenfeld, R. (2007) Stresses in two-dimensional isostatic granular systems: exact solutions. *New Journal of Physics* **9** (2007) 160–181.

Coulomb, C.A. (1785) Théorie des Machines simples, en ayant égard au frottement de leurs parties et a la roideur des Corages. Tom. X of the *Mémoires de Mathématique et de Physique Présentés à l'Académie Royale des Sciences, Par Divers Savans*, pp. 161–332.

Cundall, P.A. and Strack, O.D.L. (1979) A discrete numerical model for granular assemblies. *Géotechnique* **29** 47–65.

Dieterich, J.H. (1979) Modeling of rock friction — 1. Experimental results and constitutive equations. *Journal of Geophysical Research* **84** 2161–2168.

Dieterich, J.H. (1981) Constitutive properties of faults with simulated gouge. In: Carter, N.L., Friedman, M., Logan, J.M. and Stearns, D.W. (Eds.), *Mechanical Behavior of Crustal Rocks.*, Geophysical Monograph Series, Vol. 24. American Geophysical Union, Washington, DC, pp. 103–120.

Dorgan, K.M., Jumars, P.A., Johnson, B. and Boudray, B.P. (2006) Macrofaunal burrowing: the medium is the message. *Oceanography and Marine Biology: An Annual Review*, **44** 85–121.

Ferellec, J-F. and McDowell, G.R. (2010) A method to model realistic particle shape and inertia in DEM. *Granular Matter* **12** 459–467.

Hamaker, H.C. (1937) *The London – van der Waals Attraction between Spherical Particles. Physica* **4** (10) 1058–1072.

Hanaor, D.A.H., Gan, Y. and Einav, I. (2015) Contact mechanics of fractal surfaces by spline assisted discretisation. *International Journal of Solids and Structures* **59** 121–131.

Heyman, J. (1972) Coulomb's Memoir on Statics: An Essay in the History of Civil Engineering. CUP Archive.

Hertz, H. R. (1882) Ueber die Beruehrung elastischer Koerper *(On Contact Between Elastic Bodies),* in *Gesammelte Werke* (Collected Works), Vol. 1, Leipzig, Germany, 1895.

Hunter, R.J. (1987) *Foundations of Colloid Science*, Vol. I. Oxford: Clarendon.

Hunter, R.J. (2001) *Foundations of Colloid Science*. Oxford: OUP.

Jean, M. and Moreau, J.J. (1992) Unilaterality and dry friction in the dynamics of rigid bodies collections. *Proc. of Contact Mech. Int. Symp.* (Curnier, A. Ed) Lausanne: Presses Polytechniques et Universitaires Romandes. 31–48.

Jean, M. (1999) The non-smooth contact dynamics method. *Computer Methods in Applied Mechanics and Engineering*, **177** (3–4) 235–257.

Johnson, K.L. (1985) *Contact mechanics*, Cambridge University Press.

Koenders, M.A. and Stefanovska, E. (1993) The numerical simulation of a dense assembly of particles with elasto-frictional interaction. *Powder Technology* **77** 115–122.

Konishi, J. (1978) Microscopic model studies on the mechanical behaviour of granular material. Proc. US-Japan Seminar on Continuum Mechanical and Statistical Approaches in the Mechanics of Granular Materials, Sendai, (eds S.C. Cowin and M. Satake) pp. 27–45. Gakujutsu Bunken Fukyu-Kai, Tokyo.

Kruyt, H.R. and Overbeek, J.Th.G. (1924) Inleiding tot de fysische chemie. Amsterdam: H.J. Paris. 1924 (reprinted, 18th edition 1969).

Kuhn, M. (1999) Structured deformation in granular materials. *Mechanics of Materials* **31** 407–429.

Lambe, T.W. and Whitman, R.V. (1969) *Soil Mechanics.* New York: Wiley.

Landau, L.D. and Lifshitz, E.M. (1976). *Theory of Elasticity* (Course of Theoretical Physics, Volume 7). Oxford: Pergamon Press.

Liu, A.J. and Nagel, S.R. (2010) The jamming transition and the marginally jammed solid, *Annual Review of Condensed Matter Physics* **1** 347–369.

Macaro, G. and Utili, S. (2012) DEM Triaxial tests of a seabed sand. *Proc. Discrete Element Modelling of Particulate Media* (Wu, C-Y. ed). Cambridge: The Royal Society of Chemistry. 203–211.

McDowell, G.R. and Li, H. (2016) *Granular Matter* **18** 66–67.

Maxwell, J.C. (1864) On reciprocal figures and diagrams of forces, *Philosophical Magazine, 4th Series*, **27** 250–261.

Mindlin, R.D. (1949) Compliance of elastic bodies in contact. *J. Appl. Mech.* **16** 259–268.

Mindlin, R.D. and Deresiewicz, H. (1953) Elastic spheres in contact under varying oblique forces. *J. Appl. Mech. Trans. ASME* **20** 327–344.

Mohr, C.O. (1906) *Abhandlungen aus dem Gebiete der technischen Mechanik ; mit zahlreichen Textabbildungen.* Berlin: Wilhelm Ernst & Sohn.

Petford, N. and Koenders, M.A. (2003) Shear-induced pressure changes and seepage phenomena in a deforming porous layer, 1. *Geophysical Journal International* **155** 857–869.

Powrie, W. (2004) *Soil Mechanics: Concepts and Applications, 2nd ed.* London: Spon Press.

Puttock, M.J. and Thwaite, E.G. (1969) Elastic compression of spheres and cylinders at point and line contact. National Standards Laboratory Technical Paper No 25. Commonwealth Scientific and Industrial Research Organisation Australia, Melbourne.

Radjai, F. (2008) Contact dynamics method. *European Journal of Environmental and Civil Engineering* **12** 7–8, 871–900.

Reynolds, O. (1885) On the dilatancy of media composed of rigid particles in contact, with experimental illustrations. *Philosophical Magazine. Series 5*, **20**(127) 469–481.

Ruina, A.L. (1980) Friction laws and instabilities: a quasistatic analysis of some dry frictional behavior. Ph.D. Thesis, Division of Engineering, Brown University, Providence, Rhode Island.

Ruina, A.L. (1983) Slip instability and state variable friction laws. *Journal of Geophysical Research* **88** 10359–10370.

Terzaghi, K., Peck, R.B. and Mesri, G. (1996) *Soil Mechanics in Engineering Practice.* New York: John Wiley & Sons.

Thornton, C. Sun, G. (1993) Axisymmetric compression of 3D polydisperse systems of spheres. *Proc. Powders and Grains* (Thornton, C. Ed) Rotterdam: Balkema.

Thornton, C. (2000) Numerical simulations of deviatoric shear deformation of granular media. *Géotechnique* **50**(1) 43–53.

Thornton, C. and Antony, S.J. (1998) Quasi-static deformation of particulate media. *Phil. Trans. R. Soc. Lond. A* **356** 2763–2782.

Villagio, P. (1979) An elastic theory of Coulomb friction. *Archive for Rational Mechanics and Analysis* **70** 135–143.

Williams, J.A. and Dwyer-Joyce, R.S. (2001) Contacts between solid surfaces. Chapter 3 in *Modern Tribology Handbook*, Vol. 1, 121–162 (ed Bharat, B.) Boca Raton: CRC Press.

Yade DEM software: yade-dem.org (2019).

Chapter 2

Continuum Mechanics and Cartesian Tensor Calculus

2.1 Initial considerations

In order to describe the deformation of granular materials from a large-scale point of view it is convenient to be able to employ concepts from continuum mechanics. In this approach there is no concern for the physical constituents of the medium and their interactive properties. Continuum mechanics may be applied to all manner of systems: gases, fluids, solids, crowds, flocks of birds, etc. The fundamental concept is the *material point*, which is an element of the material that contains a large number of physical constituents and which has smooth properties. In what follows differential (infinitesimal) calculus is used and therefore some thought has to be given to the manner in which the material point is chosen. In fact, it cannot be a (mathematical) point at all. This issue is less academic than one might think. When, for example, a simulation of a granular material is set up, the number of particles must be large enough to not only capture the majority of mechanical events that will take place during the deformation process, but also to capture them in sufficient numbers to be representative of the process. This is a fraught issue in granular mechanics, as these materials are intrinsically heterogeneous and questions of spatial correlations are mostly difficult to answer. Nevertheless, it is useful to have the framework of continuum mechanics in the background.

The review of the subject given here highlights the essential topics that are useful for densely packed granular materials. There are very good general books on continuum mechanics: [Becker and Bürger, 1975], [Spencer, 1980], [Fung, 1977], [Eringen, 1989]. (Cartesian) tensor calculus is treated, for example, in the classic monograph by [Jeffreys, 1931] and in [Temple, 2004].

The medium that is treated then is assumed to be made up of material points. The deformation of the medium takes place by the material points moving with respect to one another. In what follows use will be made of *Cartesian tensors*. These are mathematical objects that describe physical quantities, such as location vectors, stresses, strains and stiffness properties. They are called Cartesian, because their properties are defined via a Cartesian coordinate system. For example, a location vector $\mathbf{x}$ has three components (x_1, x_2, x_3) in three dimensions. If a different coordinate system is used, the vector remains physically the same of course, as it still points to the same location, but the components that describe it will be different.

2.2 Rotations

One way of making a change to the coordinate system is by a rigid rotation. The transformation from a coordinate frame is a matrix called $\mathbf{Q}$. The coordinates of the vector in the rotated frame are called (y_1, y_2, y_3) and the transformation is effected by a matrix multiplication

$$\begin{pmatrix} y_1 \\ y_2 \\ y_3 \end{pmatrix} = \begin{pmatrix} Q_{11} & Q_{12} & Q_{13} \\ Q_{21} & Q_{22} & Q_{23} \\ Q_{31} & Q_{32} & Q_{33} \end{pmatrix} \begin{pmatrix} x_1 \\ x_2 \\ x_3 \end{pmatrix}$$

So, while *physically* the two vectors do the same thing, that is, point to a particular location, their *representation* is different, because the reference coordinate frame has been rotated.

The matrix notation is very cumbersome. It is easier to use the numbers of the coordinates and also number the subscripts of the matrix $\mathbf{Q}$. Then the multiplication can be written as

$$y_i = \sum_{j=1}^{3} Q_{ij} x_j$$

In fact, in most cases, people don't bother writing the summation either, noting that when a subscript appears twice, it needs to be summed. This practice is called *Einstein's summation convention*; it is incredibly useful. The matrix multiplication can then be written as follows:

$$y_i = Q_{ij} x_j$$

The rotation is a very important object and its properties need to be established. To begin with the Kronecker delta $\boldsymbol{\delta}$ is introduced

$$\delta_{ij} = 1 \text{ if } i = j$$
$$\delta_{ij} = 0 \text{ if } i \neq j$$

If the Kronecker delta is written as a matrix it takes the form of the identity

$$\boldsymbol{\delta} \to \begin{pmatrix} 1 & 0 & 0 \\ 0 & 1 & 0 \\ 0 & 0 & 1 \end{pmatrix}$$

It is therefore obvious that the inverse of a rotation, $\mathbf{Q}^{-1}$, can be obtained by solving

$$Q_{ij}^{-1} Q_{jk} = \delta_{ik}$$

The *transposed* of a matrix is denoted by a superscript T; this is merely a matter of notation: in the case of the rotation $Q_{ji}^T = Q_{ij}$.

One of the physical properties of a rotation operation is that it leaves the length of a vector invariant. Therefore, writing out the inner product of the vector $\mathbf{y}$ with itself gives

$$y_i y_i = Q_{ij} x_j Q_{ik} x_k = Q_{ki}^T Q_{ij} x_j x_k$$

This must be equal to the inner product of $\mathbf{x}$ with itself and as a result it follows that

$$Q_{ki}^T Q_{ij} = \delta_{kj}$$

So, one property that is established here is that the transposed of a rotation is its inverse.

A property that can be ascertained along the same lines concerns the determinant of a rotation. Take any three independent vectors and consider the volume that these define. Now, on rotation the vectors change representation, but their physical meaning remains the same, so the volume remains unchanged. Immediately it follows that the determinant of a rotation equals 1.

These two properties of a rotation are easily verified when a specific case is used. Take a rotation that leaves the 3-axis invariant and rotate in the 1-2-plane over an angle φ. The matrix describing this takes the form

$$\mathbf{Q}(\varphi) \to \begin{pmatrix} \cos\varphi & -\sin\varphi & 0 \\ \sin\varphi & \cos\varphi & 0 \\ 0 & 0 & 1 \end{pmatrix}$$

Indeed, it is seen that transforming back, that is rotating over an angle $-\varphi$, gives the transposed matrix and that the determinant equals unity. For small angles the rotation takes the form

$$\begin{pmatrix} 1 & -\varphi & 0 \\ \varphi & 1 & 0 \\ 0 & 0 & 1 \end{pmatrix}$$

If the rotation consists of three subsequent rotations over small angles around the three coordinate axes (the so-called *Euler angles*), then the result is

$$\begin{pmatrix} 1 & -\varphi_1 & -\varphi_2 \\ \varphi_1 & 1 & -\varphi_3 \\ \varphi_2 & \varphi_3 & 1 \end{pmatrix}$$

So that for small angles the rotation is always anti-symmetric.

2.3 The strain tensor

During the deformation of a medium a material point will move from $\mathbf{x}$ to $\mathbf{x}+\mathbf{u}$, where $\mathbf{u}$ is the displacement vector. It is desirable to have a measure for the displacement of two nearby points, in other words, what is the behaviour of $\mathbf{u}$ in the vicinity of $\mathbf{x}$. In order to achieve that, the vector $\mathbf{u}$ is expanded in a Taylor series in the point $\mathbf{x}$. Fix a coordinate system with the origin in the point $\mathbf{x}$; call the coordinates ξ_i and going up to first order leads to

$$u_i = u_i^{(0)} + \frac{\partial u_i}{\partial \xi_j} \xi_j$$

The quantity $\partial u_i / \partial \xi_j$ is called the *displacement gradient*.

For the purpose of describing the material behaviour the constant vector $\mathbf{u}^{(0)}$ is not of interest, as it basically deals with the translation of the material. Imagine a block of a material, which is carried out of the laboratory. The translation says nothing about the material behaviour. The displacement gradient is much more informative, as it describes the motion of two adjacent points in the material as they are wrenched apart. Part of the displacement gradient will be a rotation. Again, a rigid rotation says nothing about the material behaviour. The same block can be slowly rotated around and no material points have come closer or moved further apart. So, in order to arrive at a meaningful object that informs about particles of the material coming together or moving apart, the rotation has to be removed from the displacement gradient.

It is now imperative to define small deformations, the so-called *geometrical linearization limit*, which is very commonly used in the theory of the deformation of solids. This limit is valid while the components of the displacement gradient have a magnitude that is much smaller than unity. In that case the deformation gradient can be written as the sum of a symmetric and an anti-symmetric part

$$\frac{\partial u_i}{\partial \xi_j} = \frac{1}{2}\left(\frac{\partial u_i}{\partial \xi_j} + \frac{\partial u_j}{\partial \xi_i}\right) + \frac{1}{2}\left(\frac{\partial u_i}{\partial \xi_j} - \frac{\partial u_j}{\partial \xi_i}\right)$$

The second part here has the form of a rotation over small angles minus the Kronecker delta. The first part is symmetric and is called the *strain* $\mathbf{E}$

$$E_{ij} = \frac{1}{2}\left(\frac{\partial u_i}{\partial \xi_j} + \frac{\partial u_j}{\partial \xi_i}\right)$$

Consequences for the strain tensor

A number of interesting properties can be determined.

1. Any symmetrical matrix can be diagonalised by choosing a coordinate frame in which the diagonal components are just the eigenvalues. These are commonly

known as the *principal strains*; they are called $E^{(1)}$, $E^{(2)}$, $E^{(3)}$ and ordered so that $E^{(1)} \geq E^{(2)} \geq E^{(3)}$. $E^{(1)}$ is the major principal strain, $E^{(3)}$ the minor principal strain and $E^{(2)}$ the intermediate principal strain.

2. If a coordinate frame exists in which the diagonal components vanish, then that state of strain is called *pure shear*. If a rotation is added to the pure shear state, so that the whole deformation can be written as $u_1 = a_s \xi_2$ then that state is called *simple shear*.

3. The fact that there are eigenvalues means that there must be a characteristic equation. No matter what the choice of coordinate frame is, these eigenvalues are always the same; therefore, the coefficients of the characteristic equation are always the same. It follows that in d dimensions there are d eigenvalues and hence d rotation-invariant coefficients. These invariants must have physical meaning. The easiest way to see what they are is by studying a strain in a diagonal state. Calling the eigenvalues $E^{(\ell)}$, the characteristic equation is

$$\prod_{\ell=1}^{d}\left(1 - E^{(\ell)}\right) = 0$$

In two dimensions the invariants are then $E^{(1)} + E^{(2)}$ and $E^{(1)}E^{(2)}$, that is, the sum of the diagonal elements (the so-called *trace*) and the determinant.

The sum of the diagonal elements has a simple meaning. Consider a rectangle with sides of lengths L_1 and L_2, see Fig. 2.1. A deformation will change the lengths of the sides to $L_1\left(1 + E^{(1)}\right)$ and $L_2\left(1 + E^{(2)}\right)$. The area of the rectangle will therefore have changed from L_1L_2 to (neglecting products of the strain components because of geometrical linearization) $L_1L_2\left(1 + E^{(1)} + E^{(2)}\right)$; it is seen that the ratio of the change in the area to the initial area is just the trace of the deformation tensor.

Giving physical meaning to the determinant is more difficult, as this is the product of the components of the strain tensor. These are all assumed to be much smaller than unity and therefore the product, as compared to something of the order of magnitude of the trace is negligible. However, the determinant of the strain plus the Kronecker delta can be given an interpretation. Adding the Kronecker delta to the strain gives a matrix that spans two vectors $\left[\left(1 + E^{(1)}\right), 0\right]$

and $\left[0,\left(1+E^{(2)}\right)\right]$. The determinant of this matrix is of course exactly the relative change in area.

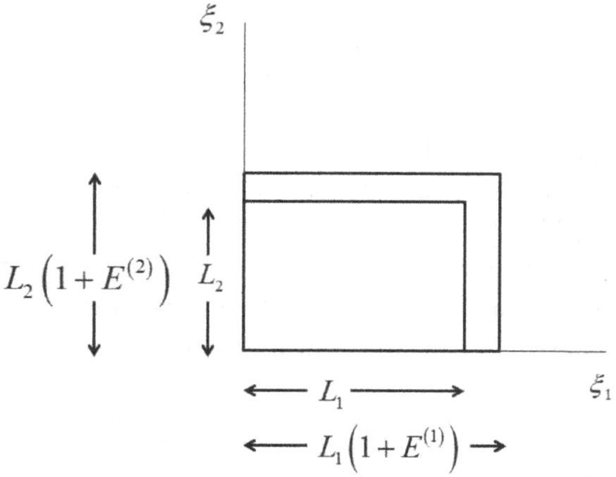

Figure 2.1. Deformation of an element of material under a strain tensor.

These results remain intact in three dimensions. The trace E_{ii} is equal to the volume strain. In three dimensions there is a third invariant, which is the sum of the sub-determinants

$$\begin{vmatrix} E_{11} & E_{12} \\ E_{21} & E_{22} \end{vmatrix} + \begin{vmatrix} E_{22} & E_{23} \\ E_{32} & E_{33} \end{vmatrix} + \begin{vmatrix} E_{11} & E_{13} \\ E_{31} & E_{33} \end{vmatrix} = E^{(1)}E^{(2)} + E^{(2)}E^{(3)} + E^{(1)}E^{(3)}$$

This invariant is used very rarely and the same thing applies to it as before: it is the product of small quantities and as a result negligible compared to the trace of the strain.

4. The extension of a line element with direction unit vector $\mathbf{n}$ in two dimensions is $n_1^2E^{(1)} + n_2^2E^{(2)}$. When $E^{(1)}$ and $E^{(2)}$ both have the same sign this describes an ellipse. When $E^{(1)}$ and $E^{(2)}$ have different signs there is a direction for which the extension vanishes. This is called the *zero-extension direction* (sometimes also referred to as the 'no-extension direction') and will be denoted by φ_{ne}.

$$\tan^2 \varphi_{ne} = -\frac{E^{(1)}}{E^{(2)}}$$

In the soil mechanics literature the solution is often represented as

$$\varphi_{ne} = \frac{\pi}{4} + \frac{1}{2}\sin^{-1}\left(\frac{E^{(1)} + E^{(2)}}{E^{(1)} - E^{(2)}}\right)$$

In fact there are four such directions; $\pm\varphi_{ne}$ and $\pm\varphi_{ne} + \pi$.

In this plane two-dimensional deformation the unit normal vector is rotated over an angle $n_1 n_2 \left(E^{(2)} - E^{(1)}\right)$.

5. The volume strain, the trace of the strain tensor, is such an important quantity that it is often quoted separately. Subtracting the volume strain from the diagonal elements of the strain, leaves a tensor, which is called the *deviatoric* part of the strain.

2.4 The stress tensor

The forces on a body in continuum mechanics are defined via the traction vector $\mathbf{T}$. The traction is the force per unit area that works on an infinitesimal area dA of the exterior of a portion of the body under consideration. The outward unit normal of the area is called $\mathbf{n}$, as illustrated in Fig. 2.2.

The stress is the decomposition of the traction on the unit normal, so that the traction is obtained; this is necessarily a two-tensor, which is called $\mathbf{\Sigma}$, in a formula

$$T_i = \Sigma_{ij} n_j$$

The force working on a finite area A is then

$$F_i = \int_A \Sigma_{ij} n_j dA$$

The stress may depend on both position and time, of course. The force on an enclosed region of the medium is

$$F_i = \oint_A \Sigma_{ij} n_j dA$$

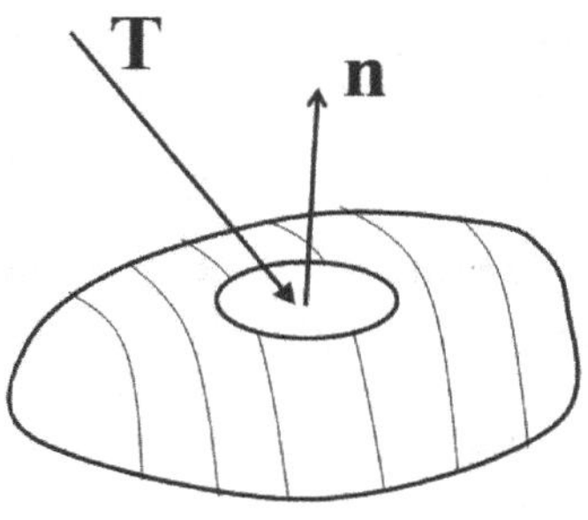

Figure 2.2. Illustration of the traction on a continuum element.

The mass of the medium in the enclosed region is the volume integral over the mass density ρ; Newton's second law in continuum terms relates the rate of change of momentum to the force acting on the medium and therefore (applying Gauss' theorem)

$$\frac{D}{Dt}\oint_V \rho v_i dV = \oint_A \Sigma_{ij} n_j dA = \oint_V \frac{\partial \Sigma_{ij}}{\partial x_j} dV$$

Here, D/Dt is the co-moving derivative and $\mathbf{v}$ the velocity. If the volume is held fixed, the integration and differentiation may be interchanged. The formula is true for arbitrary enclosed portions of the medium, which implies that the integrands of the volume integrals must be equal. Consequently, the equation of motion for a material is

$$\frac{D}{Dt}\rho v_i = \frac{\partial \Sigma_{ij}}{\partial x_j}$$

The analysis here has been concerned with forces that are transmitted to the material point via the surface. Body forces, such as gravity, have not been considered. They are easily added to the equation of motion. The gravitational force, for example, will give rise to a term ρg_i on the right-hand-side ($\mathbf{g}$ is the acceleration due to gravity).

The equation for the angular momentum requires an outer product; in index notation this is most easily done with the use of the Levi-Civita tensor $\boldsymbol{\varepsilon}$, which is defined as follows:

> $\varepsilon_{123} = 1$; if any two indices are the same, then the result is zero; any permutation of two (unequal) indices adds a minus sign.

So, for example, $\varepsilon_{132} = -1$; $\varepsilon_{231} = 1$ and $\varepsilon_{133} = 0$

Using the Levi-Civita tensor, the outer product of two vectors $\mathbf{x}$ and $\mathbf{y}$ becomes

$$(\mathbf{x} \times \mathbf{y})_i = \varepsilon_{ijk} x_j y_k$$

The angular momentum balance is now

$$\frac{D}{Dt}\oint_V \varepsilon_{ijk} x_j \rho v_k dV = \oint_A \varepsilon_{ijk} x_j \Sigma_{k\ell} n_\ell dA + \oint_V \varepsilon_{ijk} x_j \rho g_k dV$$

Again, differentiation and integration may be interchanged for fixed volumes. Noting that $\varepsilon_{ijk}\left(Dx_j / Dt\right) v_k = \varepsilon_{ijk} v_j v_k = 0$ leaves

$$\oint_V \varepsilon_{ijk} x_j \frac{D}{Dt}(\rho v_k) dV = \oint_A \varepsilon_{ijk} x_j \Sigma_{k\ell} n_\ell dA + \oint_V \varepsilon_{ijk} x_j \rho g_k dV$$

The area integral may be written as a volume integral

$$\oint_A \varepsilon_{ijk} x_j \Sigma_{k\ell} n_\ell dA = \oint_V \varepsilon_{ijk} \frac{\partial\left(x_j \Sigma_{k\ell}\right)}{\partial x_\ell} dV = \oint_V \varepsilon_{ijk}\left(\frac{\partial x_j}{\partial x_\ell}\Sigma_{k\ell} + x_j \frac{\partial \Sigma_{k\ell}}{\partial x_\ell}\right) dV$$
$$= \oint_V \varepsilon_{ijk}\left(\Sigma_{kj} + x_j \frac{\partial \Sigma_{k\ell}}{\partial x_\ell}\right) dV$$

Inserting the equation of motion cancels everything out, except for

$$\oint_V \varepsilon_{ijk} \Sigma_{kj} dV = 0$$

This must be true for arbitrary volumes, which implies that the integrand vanishes: $\varepsilon_{ijk}\Sigma_{kj}=0$. According to the rules of the Levi-Civita tensor it follows that $\Sigma_{kj}=\Sigma_{jk}$. In other words: *the stress tensor must be symmetric.*

Consequences of the symmetry of the stress tensor

The implications of the symmetry of the stress tensor are similar to those of the strain tensor.

1. The stress tensor can always be diagonalised, with eigenvalues $\Sigma^{(1)},\ \Sigma^{(2)},\ \Sigma^{(3)}$; these are called the *principal stresses*. Subtracting the mean principal stress from the stress tensor leaves the *deviatoric stress* $\Sigma_{ij}-Tr(\mathbf{\Sigma})\delta_{ij}/d$ (subtly different from the deviatoric strain, which does not include the dimension of the problem d). The term

$$p=-\frac{Tr(\mathbf{\Sigma})}{d}$$

defines the *pressure* p.

2. A graphical representation of the stress tensor is *Mohr's circle* [Mohr, 1906]. The basic concept is discussed here. To that end a two-dimensional view is taken. The stress tensor in diagonal form is

$$\begin{pmatrix} \Sigma^{(1)} & 0 \\ 0 & \Sigma^{(2)} \end{pmatrix}$$

A unit area with unit normal $\mathbf{n}$ will experience a normal force $T_i n_i$ and a tangential force $T_i\bar{n}_i$ ($\bar{\mathbf{n}}$ is the unit vector normal to $\mathbf{n}$, such that $\varepsilon_{3ij}\bar{n}_i n_j=1$). These two forces are $\frac{1}{2}\left(\Sigma^{(1)}+\Sigma^{(2)}\right)+\frac{1}{2}\left(\Sigma^{(1)}-\Sigma^{(2)}\right)\cos 2\varphi$ and $\frac{1}{2}\left(\Sigma^{(1)}-\Sigma^{(2)}\right)\sin 2\varphi$. Plotting the tangential force as a function of the normal force results in a circle with radius $\frac{1}{2}\left(\Sigma^{(1)}-\Sigma^{(2)}\right)$ and centred on $\left(\frac{1}{2}\left(\Sigma^{(1)}+\Sigma^{(2)}\right),0\right)$. This is shown in Fig. 2.3.

From the figure it is derived that the maximum ratio of tangential to normal force is achieved when the direction of the unit normal satisfies

$$\cos 2\varphi = \frac{\Sigma^{(2)} - \Sigma^{(1)}}{\Sigma^{(2)} + \Sigma^{(1)}}$$

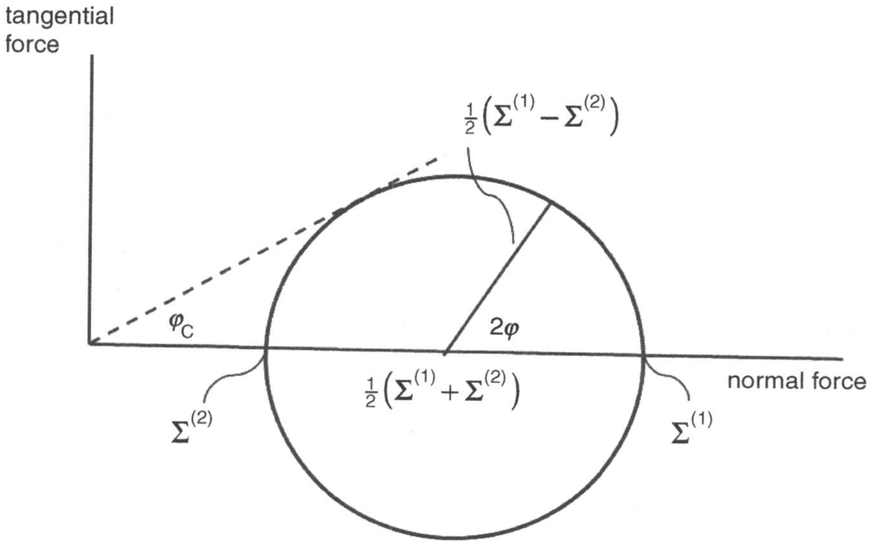

Figure 2.3. Illustration of Mohr's circle.

The angle in the plane (that is normal to the unit vector) is called the *maximum obliquity direction*. The angle φ_C that describes the maximum value of the ratio of the tangential to the normal force satisfies

$$\sin \varphi_C = \frac{\Sigma^{(1)} - \Sigma^{(2)}}{\Sigma^{(1)} + \Sigma^{(2)}}$$

These relations will be useful later on when friction phenomena are treated.

2.5 Tensors

The tensors encountered so far are the following:

- Tensors of order zero, also known as scalars. These are invariant under rotation, for example, the pressure.

- Tensors of order one, also known as vectors. For example, the displacement.
- Tensors of order two. For example, the Kronecker delta, the stress or the deformation gradient.
- Tensors of order three. For example, the Levi-Civita tensor.

The tensors describe physical quantities. When the coordinate frame is rotated, the components of the tensors change value. The number of rotation matrices required is equal to the order of the tensor. A vector $\mathbf{u}$ transforms to a rotated co-ordinate frame, denoted by an asterisk, *, as $u_i^* = Q_{ij}u_j$ under the transformation $\mathbf{Q}$ that converts the coordinates from the un-starred to the starred system. A second order tensor $\mathbf{T}$ transforms as $T_{pq}^* = Q_{pi}Q_{qj}T_{ij}$ and so on.

A scalar is the same in all coordinate frames and is therefore invariant under transformation. No (non-trivial) first order tensors are invariant, but a second order tensor that has that property is the Kronecker delta

$$\delta_{pq}^* = Q_{pi}Q_{qj}\delta_{ij} = Q_{pj}Q_{qj} = Q_{pj}Q_{jq}^T = \delta_{pq}$$

For this reason the Kronecker delta is *the identity* (tensor). This result still holds if it is multiplied by a factor. The Levi-Civita tensor is also invariant, but for a minus sign.

In what follows higher order tensors will be required, especially fourth order ones. These are discussed in the next section.

2.6 Material response

The mechanical response of a material is described by the relation between stress and strain. This rather simple statement conceals a multitude of variations and subtleties, which are obvious when one considers the enormous variety of materials and substances that exist — even in everyday life. A first order classification may be achieved by distinguishing between solids and fluids. The latter are *rate-sensitive*, which points to a connection between the stress and strain rate. Purely viscous substances, such as water or petrol, may be described in this way. However, many everyday materials defy a simple classification. Toothpaste, for example, has both fluid and solid properties.

Here the concern is with granular materials and while it may be possible to identify certain stress régimes where rate-dependence may play a role (even dry sands can exhibit 'creep', for example, and for granular assemblies of very fine particles the situation is not clear-cut) the approach taken here is that the materials are not rate-dependent. Does that mean that the relation between stress and strain is a simple proportionality? Definitely not; the relation is highly non-linear.

Capturing the non-linearity can be done by a phenomenological approach: conducting many experiments on a sample of the material and fitting the stress-strain behaviour so measured to material models, such as some version of plasticity theory. In this way an intuition, or expertise, can be built up, which can be deployed by professionals in the relevant subjects: geotechnical engineering, geophysics, chemical engineering, *etc*. The important aspect of the phenomenological approach is that there is a wealth of experimental data in the literature. It would be stupid not to make use of it. The physics of the deformation of densely packed granular materials has received not so much attention, possibly because it targets no particular application.

In a general sense the non-linearity necessitates the introduction of an approach by *increments*, as follows. At any time during the deformation the internal state of the material is noted. A small increment of strain or stress is then applied and the associated increment of stress or strain respectively is evaluated.

Incremental quantities are denoted by a small-type letter, $\boldsymbol{\sigma}$ for a stress increment and $\mathbf{e}$ for a strain increment. The question is whether the smallness of the latter was not already covered by the assumption of geometrical linearization and the answer is 'no'. Even on a range of strain that is only a few percent there can be enormous variability in the response of the material.

The incremental description enables the expression of the connection between a stress and strain increment as a linear one. So, from a given state of the material a response increment is obtained, as if the material is linear. The latter changes the internal state, leading to a new connection between stress and strain increments and then the subsequent increment is doled out. The collection of increments is called a path, so there is a *stress path* and an associated *strain path*, or a strain path and an associated stress path, depending on whether the experiment is done by stress control or strain control of a sample.

The response of the strain increment to a stress increment is termed a *compliance*. The other way round, the response of the stress increment to a strain increment, is known as a *stiffness* or *modulus*. Frequently, the word 'increment' is dropped; in the context of stress-strain relations it is always clear from the use of the symbols whether increments or total values are being discussed. It is anyway pointless to talk about the moduli or compliances when total values are in question; the relationship is, generally speaking, too non-linear for that. The total value is sometimes denoted by the prefix 'pre': *pre-stress and pre-strain.*

In general the stress-strain relationship is given by the proportionality

$$\sigma_{ij} = X_{ijk\ell}e_{k\ell} \text{ or } e_{ij} = C_{ijk\ell}\sigma_{k\ell}$$

The stiffness tensor $\mathbf{X}$ and the compliance tensor $\mathbf{C}$ are of the fourth-order. One can be obtained from the other by inverting, providing the inverse exists. Certain symmetry relations are noted. Both stress and strain are symmetric, so, for example, it follows that $X_{ijk\ell} = X_{ij\ell k} = X_{jik\ell} = X_{ji\ell k}$. Consequently, the number of independent parameters is somewhat limited. However, that still remains a lot of parameters to describe the full material response. In two dimensions there are nine parameters that need to be specified, and in three dimensions no fewer than 36. All this for each increment! The question arises whether there are any physical principles that limit the number of describing parameters.

2.7 Isotropic materials

In certain cases it can be argued that the material has no in-built direction. In that case a rotation of the co-ordinate system would leave the tensor invariant. A fourth-order tensor that is invariant under rotation has the form

$$X_{ijk\ell} = \lambda\delta_{ij}\delta_{k\ell} + \mu\delta_{ik}\delta_{j\ell} + \mu'\delta_{i\ell}\delta_{kj}$$

with three coefficients λ, μ and μ'. As $X_{ijk\ell} = X_{ij\ell k} = X_{jik\ell} = X_{ji\ell k}$, it follows that $\mu' = \mu$, no matter what the dimension is. So, the important conclusion is that *an isotropic material has two material constants.*

The constants λ and μ are called the *Lamé constants*. Elastic isotropic materials are frequently described with other sets of pairs of constants, which have direct meaning for certain applications. Young's modulus and Poisson's ratio (the contraction coefficient), commonly denoted by E and ν; the bulk and the shear modulus K and G are frequently used to characterise the material. The latter is related to the Lamé constants as $G = \mu$. The definition of the bulk modulus is the ratio of the pressure to the volume strain. All these parameters can be converted into each other. The relations are given in Appendix, Section A3. There are a lot of very good textbooks on the subject of elasticity and its very many (engineering) applications: [Landau and Lifschitz, 1976], [Timoshenko and Goodier, 1970], a very old book is [Love, 1934] and a more modern one [Barber, 2010].

The inverse of the isotropic moduli, the isotropic compliances are easily obtained by inversion. The identity of the fourth order tensors is such that $X_{ijpq}X^{-1}_{pqab} = \frac{1}{2}\left(\delta_{ai}\delta_{bj} + \delta_{aj}\delta_{bi}\right)$ (this definition preserves the symmetry in the first and second pairs of indices). The compliances take the form

$$C_{pqij} = \xi\delta_{pq}\delta_{ij} + \psi\left(\delta_{ip}\delta_{jq} + \delta_{iq}\delta_{jp}\right)$$

with $\psi = \dfrac{1}{4\mu}$ and either $\xi = -\dfrac{\lambda}{4\mu(\lambda+\mu)}$ in two dimensions, or

$$\xi = -\frac{\lambda}{2\mu(3\lambda+2\mu)} \quad \text{in three dimensions}$$

2.8 Elastic behaviour

Here a quasi-static deformation is considered. First define a function $\Im$ such that

$$\sigma_{ij} = \frac{\partial \Im}{\partial\left(\dfrac{\partial u_i}{\partial x_j}\right)}$$

Consider the deformation of the material to change from ‘state 1’ to ‘state 2’, according to some path. The work done in a small increment of deformation by the traction $\mathbf{t}$, which is associated with a surface displacement $d\mathbf{u}$ is

$$\oint_A t_p du_p dA$$

The total work done in going from state 1 to state 2 is

$$W_{12} = \int_{1}^{2} \oint_{A} t_p du_p dA$$

Now characterise the state of the material by a parameter α. State 1 corresponds to α_1 and state 2 to α_2; the changes in the state are achieved smoothly, so that the parameter α travels the interval $\alpha_1 \le \alpha \le \alpha_2$. The work done is then

$$W_{12} = \int_{\alpha_1}^{\alpha_2} \oint_{A} t_p \frac{du_p}{d\alpha} dA d\alpha = \int_{\alpha_1}^{\alpha_2} \oint_{A} \sigma_{pq} \frac{du_p}{d\alpha} n_q dA d\alpha = \int_{\alpha_1}^{\alpha_2} \oint_{V} \frac{\partial}{\xi_q} \left(\sigma_{pq} \frac{du_p}{d\alpha} \right) dV d\alpha,$$

where the definition of the stress and Gauss' theorem were used.

The material is in static equilibrium during the deformation — $\partial \sigma_{pq} / \partial \xi_q = 0$ — so the work done becomes

$$W_{12} = \int_{\alpha_1}^{\alpha_2} \oint_{V} \sigma_{pq} \frac{d\left(\partial u_p / \partial \xi_q\right)}{d\alpha} dV d\alpha$$

Using the definition of $\Im$ (and assuming that $d\Im$ is a total differential) the result is

$$W_{12} = \int_{\alpha_1}^{\alpha_2} \oint_{V} \frac{\partial \Im}{\partial\left(\partial u_p / \partial \xi_q\right)} \frac{d\left(\partial u_p / \partial \xi_q\right)}{d\alpha} dV d\alpha = \oint_{V} \left[\Im(2) - \Im(1) \right] dV$$

It follows that, if $\Im$ exists, it expresses the amount of energy per unit volume in the material. This formula also says that no matter how the path is taken the amount of energy difference between states 1 and 2 is always the same. When this is the case the material is *elastic*; returning from state 2 to state 1 involves as much work done as work returned.

An increment of work per unit volume (along the path traced by the parameter α) is equal to

$$\delta W = \sigma_{ij} \frac{\partial u_i}{\partial \xi_j}$$

Using the stiffness tensor

$$\delta W = X_{ijk\ell} \frac{\partial u_i}{\partial \xi_j} e_{k\ell}$$

Because of the symmetry in the indices i and j the only relevant part of the deformation gradient is the strain

$$\delta W = X_{ijk\ell} e_{ij} e_{k\ell}$$

Summing up, the following is found: if $\Im$ exists the deformation can be entirely characterised by the work done between two states; conversely, the only thing that physically changes during a deformation process is the work done; it follows from $\delta W = X_{ijk\ell} e_{ij} e_{k\ell}$ that the symmetry relation $X_{ijk\ell} = X_{k\ell ij}$ holds. For an isotropic material this is evidently the case and consequently, isotropic materials, for which the stiffness tensor does not change during the deformation, are elastic.

2.9 Anisotropic materials

If a material is not isotropic it is *anisotropic* (sometimes called *orthotropic*, or *aeolotropic*). It can then still be (incrementally) elastic, as long as the symmetry relation $X_{ijk\ell} = X_{k\ell ij}$ is valid. Even if a material is anisotropic there can be certain restrictive relations. For a crystalline solid, for example, the way the atoms are arranged in a regular packing (a lattice) gives rise to certain symmetry axes that are manifest in the stiffness tensor. For granular materials that are not in a regular packing this type of symmetry is not very interesting. However, the special case of transverse isotropy is important. To illustrate it, a two-dimensional approach is taken.

The fourth-order tensor is written in matrix form

$$\begin{pmatrix} \sigma_{11} \\ \sigma_{12} \\ \sigma_{22} \end{pmatrix} = \begin{pmatrix} X_{1111} & X_{1112} & X_{1122} \\ X_{1211} & X_{1212} & X_{1222} \\ X_{2211} & X_{2212} & X_{2222} \end{pmatrix} \begin{pmatrix} e_{11} \\ e_{12} \\ e_{22} \end{pmatrix}$$

The isotropic case has the form

$$\mathbf{X} \rightarrow \begin{pmatrix} \lambda + 2\mu & 0 & \lambda \\ 0 & 2\mu & 0 \\ \lambda & 0 & \lambda + 2\mu \end{pmatrix}$$

A material is said to be transverse isotropic when there exists a coordinate frame in which the stiffness tensor takes the form

$$\mathbf{X} \rightarrow \begin{pmatrix} X_{1111} & 0 & X_{1122} \\ 0 & 2\mu & 0 \\ X_{2211} & 0 & X_{2222} \end{pmatrix}$$

The symmetry axes are then the coordinate axes; the material need not possess elastic symmetries to be transverse isotropic.

2.10 Coaxiality

A peculiar property, associated with certain forms of transverse isotropy, is *coaxiality*. A material is said to be coaxial when the principal axes of the stress and strain ellipses coincide. The coordinate rotation that diagonalises the stress will also diagonalise the strain. When two matrices can be diagonalised simultaneously they commute, in other words: $\sigma_{ij}e_{jk} - e_{ij}\sigma_{jk} = 0$. Working this out for a general stiffness and requiring that the commutation holds for every combination of strain tensor components leads to the following form

$$\begin{pmatrix} X_{1111} & 2\varsigma & X_{2222} - 2\mu \\ 0 & 2\mu & 0 \\ X_{1111} - 2\mu & 2\varsigma & X_{2222} \end{pmatrix},$$

where μ and ς are arbitrary moduli.

It is seen that an isotropic material always guarantees coaxiality (as expected). However, certain anisotropic materials can also exhibit that property, providing that the off-diagonal elements satisfy the prescribed connection with the diagonals and the shear modulus. Such materials are then never elastic, because the stiffness tensor is not symmetric.

When there is a coordinate frame in which $\varsigma = 0$, it is possible for a coaxial material to be anisotropic.

The context for coaxiality is a number of papers in the soil mechanics literature in which *non-coaxial* behaviour is signalled; see, for example [Yu and Yuan, 2006].

2.11 Objectivity and pre-stressed materials, the Jaumann derivative

When any of the moduli of the incremental stress-strain relationship are of the order, or less, than the magnitude of the pre-stress a complication arises, which is related to very general physical principles. Literature on this subject is [Gurtin *et al.*, 2010] and [Dienes, 1979].

The formulation of material behaviour should not depend on the choice of co-ordinate system. Consider then two co-ordinate systems that are connected to one another by a (time-dependent) rotation $\mathbf{Q}(t)$. An incremental quantity is essentially constructed as a time derivative, multiplied by a time increment δt. The quantity starts from a certain value in a reference configuration and reaches its subsequent state after the incremental change has taken place. This state can be reached in two ways, as is demonstrated in the diagram below. The initial state in the reference configuration is situated in the top left-hand corner. The subsequent state — after the increment has been applied — is in the bottom right-hand corner. The latter can be reached either by differentiating first and then rotating, or by rotating first and then differentiating. The quantity is said to be *objective* when the two procedures arrive at the same result.

A case in point is the deformation gradient, which can be written as a symmetric and an anti-symmetric part

$$\frac{\partial u_i}{\partial \xi_j} = \frac{1}{2}\left(\frac{\partial u_i}{\partial \xi_j} + \frac{\partial u_j}{\partial \xi_i}\right) + \frac{1}{2}\left(\frac{\partial u_i}{\partial \xi_j} - \frac{\partial u_j}{\partial \xi_i}\right)$$

The time derivatives of these are studied under the rotation $\mathbf{Q}(t)$, which transforms the coordinates of the reference configuration, frame I — denoted by

$\boldsymbol{\xi}$ — to frame II, called $\mathbf{x}$; kinematic variables in this frame are distinguished by an asterisk. The deformation velocity is called $\mathbf{v}$.

Expressing the two elements of the deformation gradient velocity in frame II is done as follows

$$\frac{1}{2}\left(\frac{\partial v_i^*}{\partial x_j} \pm \frac{\partial v_j^*}{\partial x_i}\right) = \frac{1}{2}\left(\frac{\partial v_i^*}{\partial \xi_k}\frac{\partial \xi_k}{\partial x_j} \pm \frac{\partial v_j^*}{\partial \xi_k}\frac{\partial \xi_k}{\partial x_i}\right) =$$

$$\frac{1}{2}\left(\frac{\partial}{\partial \xi_k}\left(\frac{\partial Q_{i\ell}}{\partial t}\xi_\ell + Q_{i\ell}v_\ell\right)\frac{\partial \xi_k}{\partial x_j} \pm \frac{\partial}{\partial \xi_k}\left(\frac{\partial Q_{j\ell}}{\partial t}\xi_\ell + Q_{j\ell}v_\ell\right)\frac{\partial \xi_k}{\partial x_i}\right)$$

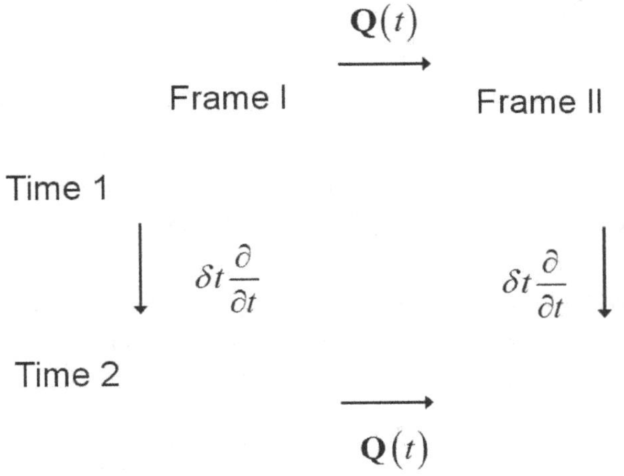

$\boldsymbol{\xi}$ is related to $\mathbf{x}$ by the rotation: $x_i = Q_{ij}\xi_j$ and so $\partial x_i / \partial \xi_j = Q_{ij}$ (also $\partial \xi_i / \partial x_j = Q_{ij}^{-1} = Q_{ji}$) and the rotation is rigid and therefore position-independent. Using these properties of $\mathbf{Q}$ and $\partial \xi_\ell / \partial \xi_k = \delta_{\ell k}$ then yields

$$\frac{1}{2}\left(\frac{\partial v_i^*}{\partial x_j} \pm \frac{\partial v_j^*}{\partial x_i}\right) = \frac{1}{2}\left(\left(\frac{\partial Q_{ik}}{\partial t} + Q_{i\ell}\frac{\partial v_\ell}{\partial \xi_k}\right)Q_{jk} \pm \left(\frac{\partial Q_{jk}}{\partial t} + Q_{j\ell}\frac{\partial v_\ell}{\partial \xi_k}\right)Q_{ik}\right)$$

Now, for the case of the plus sign, two terms combine

$$\frac{\partial Q_{ik}}{\partial t}Q_{jk} + \frac{\partial Q_{jk}}{\partial t}Q_{ik} = \frac{\partial}{\partial t}\left(Q_{ik}Q_{jk}\right)\frac{\partial}{\partial t}\left(Q_{ik}Q_{kj}^{-1}\right) = 0$$

It follows that

$$\frac{1}{2}\left(\frac{\partial v_i^*}{\partial x_j}+\frac{\partial v_j^*}{\partial x_i}\right)=\frac{1}{2}\left(Q_{jk}Q_{i\ell}\frac{\partial v_\ell}{\partial \xi_k}+Q_{ik}Q_{j\ell}\frac{\partial v_\ell}{\partial \xi_k}\right)$$

This is exactly the transformation for a second order tensor under a rotation, as expected. Therefore, the strain rate is objective.

For the case of minus sign, however, relating to the rate of change of the rotational part of the deformation gradient, denoted by $\dot{\mathbf{R}}$, it is found that

$$\dot{R}_{ij}^* = \frac{1}{2}\left(\frac{\partial v_i^*}{\partial x_j}-\frac{\partial v_j^*}{\partial x_i}\right)=\frac{1}{2}\left(\frac{\partial Q_{ik}}{\partial t}Q_{jk}-\frac{\partial Q_{jk}}{\partial t}Q_{ik}\right)+Q_{i\ell}Q_{jk}\dot{R}_{\ell k}$$

$$=Q_{i\ell}Q_{jk}\dot{R}_{\ell k}-\frac{\partial Q_{jk}}{\partial t}Q_{ik}$$

This is not the transformation for a second order tensor; an extra term has to be added. The rotation part of the deformation gradient is therefore not objective.

The issue of objectivity acquires a certain poignancy when the Cauchy stress is investigated. The transformation takes the form in compact notation

$$\dot{\mathbf{\Sigma}}^* = \dot{\mathbf{Q}}\mathbf{\Sigma}\mathbf{Q}^{-1}+\mathbf{Q}\dot{\mathbf{\Sigma}}\mathbf{Q}^{-1}+\mathbf{Q}\mathbf{\Sigma}\dot{\mathbf{Q}}^{-1}$$

From

$$\dot{R}_{ij}^* = Q_{i\ell}Q_{jk}\dot{R}_{\ell k}-\frac{\partial Q_{jk}}{\partial t}Q_{ik} \text{ or } \dot{\mathbf{R}}^* = \mathbf{Q}\dot{\mathbf{R}}\mathbf{Q}^{-1}-\mathbf{Q}\frac{\partial \mathbf{Q}^{-1}}{\partial t},$$

it follows that

$$\frac{\partial \mathbf{Q}}{\partial t} = \dot{\mathbf{R}}^*\mathbf{Q}-\mathbf{Q}\dot{\mathbf{R}}$$

This is used to evaluate the Cauchy stress rate, with the result

$$\dot{\mathbf{\Sigma}}^* = \left(\dot{\mathbf{R}}^*\mathbf{Q}-\mathbf{Q}\dot{\mathbf{R}}\right)\mathbf{\Sigma}\mathbf{Q}^{-1}+\mathbf{Q}\dot{\mathbf{\Sigma}}\mathbf{Q}^{-1}+\mathbf{Q}\mathbf{\Sigma}\left(-\mathbf{Q}^{-1}\dot{\mathbf{R}}^*+\dot{\mathbf{R}}\mathbf{Q}^{-1}\right)$$

Rearranging leads to the form

$$\dot{\mathbf{\Sigma}}^*-\dot{\mathbf{R}}^*\mathbf{\Sigma}^*+\mathbf{\Sigma}^*\dot{\mathbf{R}}^*=\mathbf{Q}\left(\dot{\mathbf{\Sigma}}-\dot{\mathbf{R}}\mathbf{\Sigma}+\mathbf{\Sigma}\dot{\mathbf{R}}\right)\mathbf{Q}^{-1}$$

It is seen that the time derivative

$$\overset{\nabla}{\mathbf{\Sigma}}=\dot{\mathbf{\Sigma}}-\dot{\mathbf{R}}\mathbf{\Sigma}+\mathbf{\Sigma}\dot{\mathbf{R}}$$

is objective and it is called the *Jaumann derivative*, or sometimes the *co-rotational derivative*. Note that the Cauchy stress itself has no objective time derivative.

Material behaviour should not depend on the rate of rotation and therefore the appropriate way to describe incremental behaviour is to specify a link between the Jaumann increment and the strain increment, which are both objective measures. The difference between the Jaumann derivative and the direct derivative of the Cauchy stress to obtain a stress increment is negligible while the incremental stiffness moduli are much greater than the pre-stress. In certain cases, however the difference is important, as will be demonstrated below.

The Jaumann derivative is symmetric.

2.12 Frictional materials

A material is said to be (pure) frictional when there are no combinations of incremental strain components that increase the stress ratio. A pre-stress state is put forward in diagonal form

$$\begin{pmatrix} P_1 & 0 \\ 0 & P_2 \end{pmatrix}$$

Added to this is an increment of stress

$$\begin{pmatrix} P_1+\sigma_{11} & \sigma_{12} \\ \sigma_{12} & P_2+\sigma_{22} \end{pmatrix}$$

The principal stresses of this new stress state are

$$\begin{pmatrix} P_1' & 0 \\ 0 & P_2' \end{pmatrix}$$

The changes in the principal stresses are evaluated up to first order in the increments

$$P_1' = P_1 + \sigma_{11} + O\left(|\boldsymbol{\sigma}|^2\right) \text{ and } P_2' = P_2 + \sigma_{22} + O\left(|\boldsymbol{\sigma}|^2\right)$$

The initial stress ratio is $\left|\frac{P_1 - P_2}{P_1 + P_2}\right|$, the one after a stress increment is $\left|\frac{P_1' - P_2'}{P_1' + P_2'}\right|$. The two are required to be equal, so (again up to first order in the increments) it follows that

$$\frac{P_2\sigma_{11} - P_1\sigma_{22}}{\left(P_1 + P_2\right)^2} = 0$$

Here, $P_1 > P_2 > 0$ has been chosen. Substituting for the stress increments the stress-strain relation results in

$$\left(P_1 X_{2211} - P_2 X_{1111}\right)e_{11} + 2\left(P_1 X_{2221} - P_2 X_{1121}\right)e_{12} + \left(P_1 X_{2222} - P_2 X_{1122}\right)e_{22} = 0$$

This must hold for arbitrary strain increments, leading to a form for the stiffness tensor as follows

$$\begin{pmatrix} X_{1111} & \varsigma & \frac{P_1}{P_2} X_{2222} \\ X_{1211} & 2\mu & X_{1222} \\ \frac{P_2}{P_1} X_{1111} & \frac{P_2}{P_1}\varsigma & X_{2222} \end{pmatrix}$$

The friction criterion makes no statement about the value of the shear modulus or the coupling between shear stress and diagonal strains. (No insight is lost if the latter are set to zero: $X_{1211} = X_{1222} = \varsigma = 0$.) It is noticed that the stiffness has no inverse; therefore, a compliance formulation is not available for a frictional material. A better understanding of that and its implications will be presented in the next section. The frictional material has a transverse anisotropic form.

References

Barber, J.R. (2010) *Elasticity*. Dordrecht: Springer.

Becker, E. and Bürger, W. (1975) *Kontinuumsmechanik*. Stuttgart: B.G. Teubner.

Dienes, J. (1979) On the analysis of rotation and stress rate in deforming bodies. *Acta Mechanica* **32**(4) 217–232.

Eringen, A.C. (1989) *Mechanics of Continua*. Malabar, Fla.: R.E. Krieger Pub. Co.

Fung, Y.C. (1977) *A First Course in Continuum Mechanics*, 2nd Edition. Englewood Cliffs: Prentice Hall.

Gurtin, M.E., Fried, E. and Anand, L. (2010) *The Mechanics and Thermodynamics of Continua*. Cambridge: CUP.

Jeffreys, H. (1931) *Cartesian Tensors*. Cambridge: Cambridge University Press.

Landau, L.D. and Lifshitz, E.M. (1976) *Theory of Elasticity* (Course of Theoretical Physics, Volume 7). Oxford: Pergamon Press.

Love, A.E.H. (1934) *A Treatise on the Mathematical Theory of Elasticity*. Cambridge: CUP.

Mohr, C.O. (1906) *Abhandlungen aus dem Gebiete der technischen Mechanik ; mit zahlreichen Textabbildungen*. Berlin: Wilhelm Ernst & Sohn.

Spencer, A.J.M. (1980) *Continuum Mechanics*. London: Longman.

Temple, G. (2004) *Cartesian Tensors: An Introduction*. Mineola (NY): Dover.

Timoshenko, S. and Goodier, J.N. (1970) *Theory of Elasticity*. New York: McGraw-Hill.

Yu, H.S. and Yuan, X. (2006) On a class of non-coaxial plasticity models for granular soils. *Proceedings of the Royal Society London, Series A* **462** 725–748.

Chapter 3

The Bounds of Static Equilibrium

3.1 Criteria for rupture

Forms of the stiffness or compliance tensor may be suggested by physical principles, such as have been encountered in the previous chapter. For a stiffness tensor to make sense, however, it must lead to equilibrium equations that can be solved (given sufficient boundary conditions). The equilibrium equations are here pursued in the absence of body forces. Also, the difference between the Jaumann stress increment and the Cauchy stress increment is not considered in the first instance, though that will be reintroduced at a later stage.

As seen before, the equilibrium equations take the form

$$\frac{\partial \sigma_{ij}}{\partial x_j} = 0 \rightarrow \frac{\partial}{\partial x_j}\left(X_{ijk\ell} e_{k\ell}\right) = 0$$

The situation of primary interest is the one in which a homogeneous material is stressed uniformly. The solution in that case is simply $u_i = \alpha_{ij} x_j$; but is that the whole solution, or are there other ones? Here, the possibility of a strongly localised displacement field is studied. Such fields are called *rupture layers* (sometimes, slip bands or failure planes).

The analysis using a stiffness-based approach was first introduced by [Biot, 1965], see also [Hill and Hutchinson, 1975]. Many authors in the soil mechanics literature have written papers on localisation that use a compliance-based model, because that makes a connection with the plasticity constitutive material description that is frequently used. Pioneer in this area has been Vardoulakis: [Vardoulakis *et al.*, 1978], [Vardoulakis, 1979, 1980], [Vardoulakis and Sulem, 1995]. There are other contributors to the theory, such as [Vermeer, 1990] whose analysis is an adaptation of the work by [Rudnicki and Rice, 1975]. Experimental work on rupture layer formation has been a key driver of this subject. Work by [Arthur *et al.*, 1977], [Arthur and Dunstan, 1982], and [Desrues and Viggiani, 2004] have shown that there are a number of possibilities for localised deformation in frictional materials and it is the task of theoreticians to produce the appropriate modelling to describe it.

The mathematical work leans heavily on the theory of second order partial differential equations. Now, it is interesting to note that the theoretical work on rupture layer formation, quoted above, generally takes the actual equations into account only (frequently, the rupture layer formation is seen as a way to establish certain aspects of the constitutive equations); however, *boundary conditions* should be specified as well and below that aspect is treated further.

To begin with, a simple analysis with an assumed form of the constitutive response is carried out. The localised layer has a unit normal $\mathbf{n}$ (the treatment is sufficient in two dimensions). The field is constant along the straight lines as shown in Fig. 3.1. In equilibrium the requirement for the rupture layer field $\mathbf{g}(\mathbf{x}.\mathbf{n})$ is

$$X_{ijk\ell} g_k'' n_j n_\ell = 0\,, \tag{3.1}$$

which has a solution when

$$\det\left(X_{ijk\ell} n_j n_\ell\right) = 0$$

The tensor $P_{ik} = X_{ijk\ell} n_j n_\ell$ is sometimes called the *acoustic tensor*.

An incremental transverse isotropic material model is used, see Section 2.9. The determinant may be worked out to give the following equation:

$$\mu X_{1111} n_1^4 + \left[\left(X_{1111} X_{2222} - X_{1122} X_{2211}\right) - \mu\left(X_{2211} + X_{1122}\right)\right] n_1^2 n_2^2 \\ + \mu X_{2222} n_2^4 = 0$$

The material will be able to experience a rupture layer when there are real solutions for n_1 and n_2, or, alternatively dividing through by n_1^4, for the tangent of the angle that the normal of the rupture layer makes with the x_1 axis, which will be called $\sqrt{y}$. The equation for y reads then

$$ay^2 + by + c = 0,$$

$$\text{with } a = \mu X_{2222}\,;\ b = \left(X_{1111}X_{2222} - X_{1122}X_{2211}\right) - \mu\left(X_{2211} + X_{1122}\right);$$
$$c = \mu X_{1111}$$

The solution is simply

$$y = \frac{-b \pm \sqrt{b^2 - 4ac}}{2a}$$

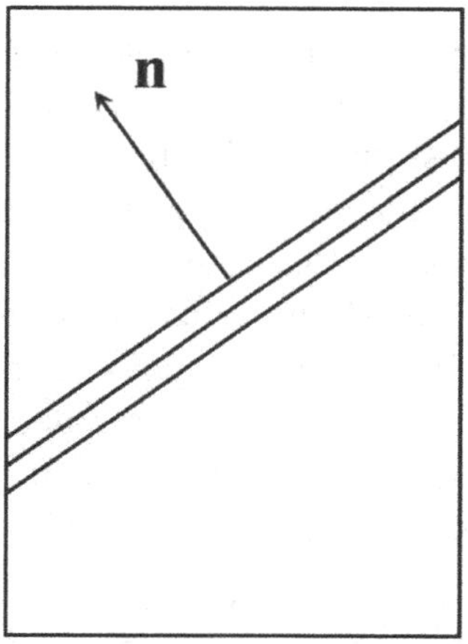

Figure 3.1. Illustration of the geometry of rupture layer formation.

The solution must be real positive for the rupture phenomenon to exist.

At this point it is helpful to outline a scenario. A sample of granular material is stressed, increment after increment; as a result the incremental moduli will change. If the loading is such that — after an initial isotropic compression — the stress Σ_{11} increases, while Σ_{22} is held constant, then the sample will come to a point at which peak stress is reached. Rupture phenomena are usually observed at, or near, this point. A dense sample will dilate and, taking compressive stresses and strains positive, as is convention in soil mechanics, $e_{22} > -e_{11} < 0$. At peak stress the increment $\sigma_{11} = 0$. In a test in which Σ_{22} is held constant, so that the increment $\sigma_{22} = 0$, implies that

$$\frac{X_{1111}X_{2222} - X_{1122}X_{2211}}{X_{2222}} = 0$$

In other words, at peak stress the 'outer determinant' of the stiffness tensor vanishes. The values of the major and minor principal stress at this point will be called P_1 and P_2.

It is important to have an impression of how the outer determinant approaches zero. At a point just before peak, the major principal stress will have reached a value that is a small amount smaller than the peak value, $P_1(1-\varepsilon)$ say. In order to arrive at peak the principal strain must be increased by a certain amount, Δe_{11}. Therefore, the ratio $\varepsilon P_1 / \Delta e_{11}$ gives an impression of the stiffness. The ratio $\varepsilon / \Delta e_{11}$ is a number that approaches zero at peak, but at some point before peak could easily be of the order of unity (for example, at 99% of the stress peak it will take another 1% of strain to reach the actual peak value). Consequently, the stiffness near peak is of the order of magnitude of the peak stress, say fP_1, where the number f may vary from the order of unity to zero (and in principle it could be negative, when post-peak behaviour is included). For the purposes of a calculation it is then practical to set

$$\frac{X_{1111}X_{2222} - X_{1122}X_{2211}}{X_{2222}} = fP_1$$

In the scenario sketched, where the moduli change continuously as the stress ratio is pushed up, the point at which the rupture layer materialises is when $b^2 - 4ac$ becomes zero (while $-b/2a$ is positive). Now,

$$
\begin{aligned}
&b^2 - 4ac = \\
&\left[\left(X_{1111}X_{2222} - X_{1122}X_{2211}\right) - \mu\left(X_{2211} + X_{1122}\right)\right]^2 - 4\mu^2 X_{1111}X_{2222} \\
&= \left[fP_1X_{2222} - \mu\left(X_{2211} + X_{1122}\right)\right]^2 - 4\mu^2 X_{1111}X_{2222} \\
&= f^2P_1^2X_{2222}^2 - 2fP_1\mu X_{2222}\left(X_{2211} + X_{1122}\right) \\
&+\mu^2\left(X_{2211} + X_{1122}\right)^2 - 4\mu^2 X_{1111}X_{2222}
\end{aligned}
$$

This becomes zero at the point when the shear modulus reaches the value

$$
\begin{aligned}
\mu &= \frac{fP_1X_{2222}}{X_{1122} + X_{2211} \pm \sqrt{X_{1111}X_{2222}}} \\
&= \frac{fP_1X_{2222}}{X_{1122} + X_{2211} \pm \sqrt{fP_1X_{2222} + X_{1122} + X_{2211}}}
\end{aligned}
$$

The direction of the rupture layer for these solutions are given by

$$
-\frac{b}{2a} = \pm\frac{\sqrt{fP_1X_{2222} + X_{1122}X_{2211}}}{X_{2222}}
$$

Some practical input is required to assess the meaning of these findings. In a test in which the minor principal stress is kept constant the ratio $X_{2211} / X_{2222} = -e_{22} / e_{11}$, which is the dilatancy ratio δ. In the previous chapter — Section 2.11 — it was shown that for a material in a frictional state $X_{1122} / X_{2222} = P_1 / P_2$, which is the principal stress ratio R. Both δ and R are of the order of magnitude of unity. The moduli X_{1111}, X_{1122}, X_{2211}, X_{2222} are all very much greater than the principal stresses; typically $X_{\bullet} \simeq 1000P_1$. Using these numbers the shear modulus at the point of incipient rupture layer formation is of the order of

$$
\mu \simeq fP_1
$$

As it was argued that f is of the order of unity, leading to zero for peak stress conditions, it is observed that a shear modulus of the same order of magnitude as the pre-stress needs be to taken into account. It was therefore wrong to ignore the difference between the Jaumann stress increment and the Cauchy stress increment. The analysis needs to be done again, using the Jaumann derivative.

The Jaumann incremental formulation ensures that the material model is phrased in a way that it moves with the material. Thus, the stiffness tensor $\mathbf{X}$ connects the Jaumann increment with the strain increment; both are objective. Equilibrium, however, is phrased using the Cauchy stress increment. The connection between the Cauchy stress increment and the Jaumann stress increment has been established in Section 2.11.

$$\dot{\mathbf{\Sigma}} = \overset{\nabla}{\mathbf{\Sigma}} + \dot{\mathbf{R}}\mathbf{\Sigma} - \mathbf{\Sigma}\dot{\mathbf{R}} \rightarrow \sigma_{ij} = X_{ijk\ell}e_{k\ell} + r_{ik}P_{kj} - P_{ik}r_{kj} \text{ and } \frac{\partial \sigma_{ij}}{\partial x_j} = 0$$

The system of equations for the disturbance $\mathbf{g}$ is then somewhat modified with terms that contain the pre-stress

$$\begin{pmatrix} n_1^2 X_{1111} + n_2^2\left[\mu - \tfrac{1}{2}(P_1 - P_2)\right] & n_1 n_2\left[X_{1122} + \mu + \tfrac{1}{2}(P_1 - P_2)\right] \\ n_1 n_2\left[X_{2211} + \mu - \tfrac{1}{2}(P_1 - P_2)\right] & n_2^2 X_{2222} + n_1^2\left[\mu + \tfrac{1}{2}(P_1 - P_2)\right] \end{pmatrix}\begin{pmatrix} g_1'' \\ g_2'' \end{pmatrix} = 0$$

The solution becomes possible when the determinant of the matrix vanishes. As before, write for the tangent of the angle that the normal of the rupture layer makes with the x_1 axis $\sqrt{y}$, then

$$ay^2 + by + c = 0,$$

with

$$a = \left[\mu - \tfrac{1}{2}(P_1 - P_2)\right]X_{2222};$$
$$b = (X_{1111}X_{2222} - X_{1122}X_{2211}) - \mu(X_{2211} + X_{1122}) - \tfrac{1}{2}(P_1 - P_2)(X_{2211} - X_{1122});$$
$$c = \left[\mu + \tfrac{1}{2}(P_1 - P_2)\right]X_{1111}$$

The solution for y is

$$y = \frac{-b \pm \sqrt{b^2 - 4ac}}{2a}$$

Onset of the rupture layer occurs at the point when the discriminant flips from negative to positive, while $-b/(2a)$ is positive. A substantial amount of algebra is required to obtain the solution. The following abbreviations are employed (an approximately frictional material is envisaged)

$$R = \frac{X_{1122}}{X_{2222}} = \frac{P_1}{P_2};\ \delta = \frac{X_{2211}}{X_{2222}}$$

The solution for y is

$$y = \frac{2\mu(R+\delta) - P_2\left(R^2 + R(2f-\delta-1)+\delta\right)}{2\left[2\mu - P_2(R-1)\right]}$$
$$\pm \frac{\sqrt{X_{2222}\left(P_2^2 D_1 - 4\mu P_2 D_2 + 4\mu^2 D_3\right) + 4P_2Rf\left(P_2(R-1)+2\mu\right)\left(P_2(R-1)-2\mu\right)}}{2\sqrt{X_{2222}}\left[2\mu - P_2(R-1)\right]}$$

where

$$\begin{aligned}
D_1 &= R^4 + 2R^3(2f+\delta-1) + R^2\left(4f^2 - 4f(\delta+1) + \delta^2 - 4\delta + 1\right) \\
&\quad + 2R\delta(2f-\delta+1) + \delta^2 \\
D_2 &= R^3 + R^2(2f-1) + R\delta(2f-\delta) + \delta^2 \\
D_3 &= (R-\delta)^2
\end{aligned}$$

Incipient rupture becomes possible when the term under the square root vanishes, which takes place when the shear modulus equals

$$\frac{\mu^{\pm}}{P_2} = \frac{X_{2222}D_2}{2\left(X_{2222}(R-\delta)^2 - 4P_2Rf\right)}$$
$$\pm \frac{2R\sqrt{f}\sqrt{\left[\left(X_{2222}\delta + P_2 f\right)\left(X_{2222}\left(R^2 + R(f-\delta-1)+\delta\right) + P_2(R-1)^2\right)\right]}}{X_{2222}(R-\delta)^2 - 4P_2Rf}$$

In order to illustrate these values a plot has been made for a choice of sample values (Fig. 3.2): $X_{2222} = 10^7\,Nm^{-2}$; $P_2 = 10^{-3} X_{2222}$; $R = 5$; $\delta = 2$. These values correspond roughly to those of a dense sand near peak stress ratio.

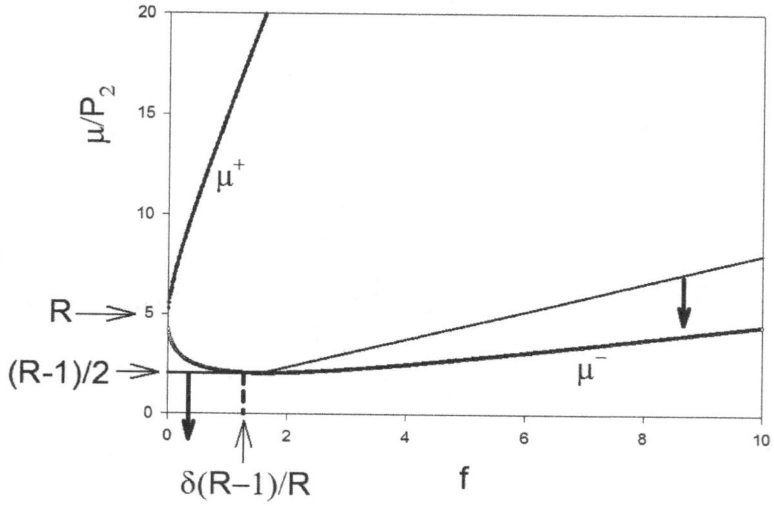

Figure 3.2. Values for the moduli at transition to rupture.

The region where $-b/a$ is negative is marked by the bold-type arrows. For $f > \delta(R-1)/R$ this region is represented by the sloping straight line; for $f \le \delta(R-1)/R$ it is the flat plateau at $\mu/P_2 = (R-1)/2$. Transitions to rupture that lead to positive values of $y = -b/a$ are of interest only. In the graph the region between the two lines μ^+ and μ^- represents a negative discriminant. Transitions can therefore take place at the line μ^+ for $f \ge 0$ and at the line μ^- in the range $0 \le f < \delta(R-1)/R$. Question is then: what values of $y = -b/a$ correspond to these two cases. The graph shown in Fig. 3.3 provides the answer.

It is noted that the values for μ/P_2 on transition at the line μ^+ increase rapidly with increasing values of f. When μ/P_2 is large the objection against the original analysis, in which the Jaumannian increment was not considered, becomes invalid. It is then reasonable to estimate the value of y that belongs to this case. One obtains in the limit that the pre-stress is much smaller than the main moduli

$$-\frac{b}{2a}=\frac{\sqrt{fP_1X_{2222}+X_{1122}X_{2211}}}{X_{2222}}\rightarrow\sqrt{\delta R}$$

The latter value is plotted as the dashed line in the graph of y as a function of f. It is the bottom curve in this graph and belongs to the transition to rupture at the line μ^+ in Fig. 3.2.

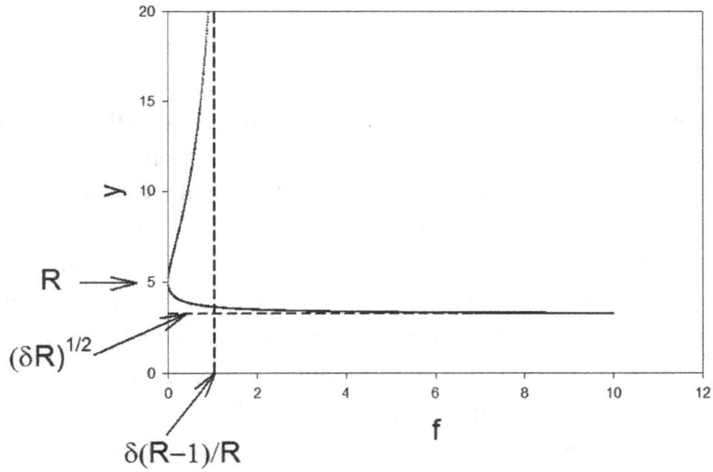

Figure 3.3. The direction of the tangent of the rupture layer direction squared, y, as a function of f.

The transition at μ^- leads to a whole range of values for y, including very large ones, corresponding to $n_2/n_1\rightarrow\infty$. In this case the rupture layer is aligned with the major principal stress axis.

3.2 The context of second order, partial differential equations

The question now is: what will actually happen at the point of transition? This depends on the boundary conditions that apply. An overview of the role of boundary conditions in second order linear partial differential equations is given in [Morse and Feshbach, 1953]. The overview hinges on the classification of these equations, which is done as follows:

If $b^2 - 4ac < 0$, equation (3.1) is said to be *elliptic*

If $b^2 - 4ac = 0$ equation (3.1) is said to be *parabolic*

If $b^2 - 4ac > 0$ equation (3.1) is said to be *hyperbolic*

The boundary conditions that are applied must be such that the solution is stable, which — quoting [Jackson, 1962] — implies that 'A stable solution is one for which small changes in the boundary conditions cause appreciable changes in the solution only in the neighbourhood of the boundary'. [Morse and Feshbach, 1953] summarise in Table 3.1 the requirements on the boundary conditions for these types of equations.

The transition from elliptical to hyperbolic is called a *bifurcation*. The vast majority of tests in which rupture layers are observed are conducted in such a way that there is a closed boundary. That is certainly the aim of the experiment, whether it be done under stress control or strain control. At the point of bifurcation the stable, unique solution is abandoned. If the medium goes into the hyperbolic mode the apparatus over-specifies the boundary conditions. Therefore, such tests can only be done in an imperfect apparatus. Now, every apparatus has imperfections when it comes to specifying boundary conditions. It is impossible to say, for example, what exactly is specified near the corners. While the material is still elliptical that does not matter so much, because the influence of an anomalous specification is only felt near the corner, see Jackson's quote, above. Rupture layers frequently emanate from the corners of an apparatus.

Table 3.1. Summary of requirements on the boundary for different types of linear second order partial differential equations.

Type of boundary conditions	*Elliptic equation*	*Hyperbolic equation*	*Parabolic equation*
Dirichlet open surface (displacement specified)	Not enough	Not enough	Unique, stable solution in one direction
Dirichlet closed surface (displacement specified)	**Unique, stable solution**	Too much	Too much
Neumann open surface (displacement gradient specified)	Not enough	Not enough	Unique, stable solution in one direction
Neumann closed surface (displacement gradient specified)	**Unique, stable solution**	Too much	Too much
Cauchy open surface (both displacement and displacement gradient specified)	Unphysical results	**Unique, stable solution**	Too much
Cauchy closed surface (both displacement and displacement gradient specified)	Too much	Too much	Too much

The analysis suggests that certain rupture layers are possible, but the continuum is not compelled to take the rupture route. So, the possibilities are as follows.

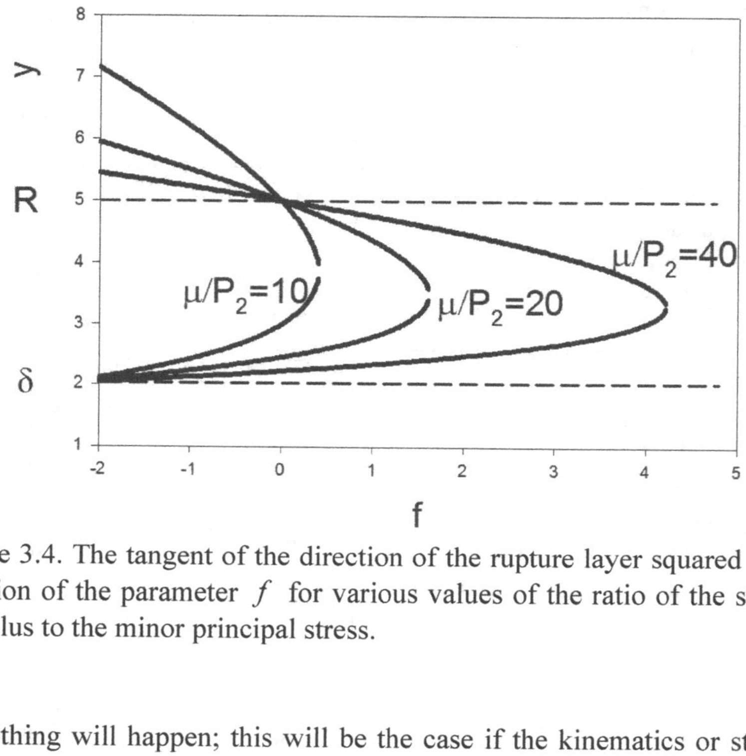

Figure 3.4. The tangent of the direction of the rupture layer squared as a function of the parameter f for various values of the ratio of the shear modulus to the minor principal stress.

1. Nothing will happen; this will be the case if the kinematics or stress does not comply with any boundary conditions. The moduli cross over into territory where rupture may occur; this territory may be termed the *metastable régime*. However, if they do so — for example, if they cross the line μ^+, two possible directions present themselves, while the original direction of a rupture layer is lost. This development has been plotted in Fig. 3.4 for various values of μ / P_2. The initial direction of rupture is to the right (where only one direction is possible), but then as f is made smaller two directions that satisfy the equations appear. It is seen that these two directions share a common point, no matter what the

value of μ / P_2 is, for $y = R$ (at $f = 0$) and $y = \delta$ (at $f = -\delta$); these are the maximum obliquity and no-extension directions. Note that to reach $y = \delta$ the value of f must be negative, that is, the assembly is in a post-peak state. The direction $y = \sqrt{R\delta}$ corresponds approximately to the intermediate direction, that is the value of the angle halfway between the maximum obliquity and no-extension directions.

The actual angles are as follows:

$$\text{No extension direction: } \tan^2 \varphi_{ne} = -\frac{e_{11}}{e_{22}}$$

$$\text{Maximum obliquity direction: } \tan^2 \varphi_{mo} = \frac{P_1}{P_2}$$

Once the medium is in the meta-stable regime the equilibrium equations are no longer elliptical; the problem becomes hyperbolic and motion takes place along characteristics.

2. A second possibility on reaching the rupture criterion is that the continuum is able to make internal changes in such a way that the moduli perambulate down the transition line μ^+ until they reach the point at which they can go no further and reach $f = 0$ at which point the maximum obliquity direction is invoked. Alternatively, this mechanism becomes exhausted somewhere along the way and rupture takes place at an angle between maximum obliquity and intermediate. While this mechanism is speculative, it is plausible on the basis of the fact that the media considered are very heterogeneous and it is therefore possible that locally the rupture criterion is achieved, while the medium as a whole is not yet there. The mechanism thus implies quite non-uniform motion.

3. On reaching the rupture criterion, the material is covered in a lattice network — not necessarily regular — of rupture planes along which deformation takes place. A new continuum is then created, which is lumpy. Models for post-rupture behaviour have been proposed that employ this philosophy. They are called *double-sliding* models. Note that due to their lumpiness an internal length scale is required. Such materials permit two types of rupture: one that takes place along the original rupture directions and one that encompasses many lumps. The directions

of these two types of rupture may be different (but generally they are the main directions already identified). Double sliding models introduce complications. One is that the lumps themselves are usually taken to be rigid and as a result it is not immediately clear how the internal rotation is determined during any incremental deformation. Various suggestions have been put forward, from 'free' rotation (within certain limits) to rotation that follows the rotation of the stress increment, see [De Josselin de Jong, 1977] for the former and [Spencer, 1964] for the latter.

Generally speaking the scenario sketched here of moduli that change slowly as the material reaches peak stress (ratio) and has frictional/dilational properties is likely to be correct. Once the deformation becomes non-uniform, as the moduli cross the rupture criterion, 'traditional' continuum mechanics is probably insufficient to describe what takes place. Also, treating the continuum as a self-contained entity is no longer correct; the boundary conditions must be taken into account.

There has been a lot of effort matching the directions of the rupture layers to experimental observations. The experimental study has been most successful using X-rays in which the rupture shows up as a less materially dense (increased dilatancy) narrow band. The angles of the layers with respect to the major principal stress direction bunch around the three main angles obtained in the theory. Using an apparatus with rubber boundaries, which possibly imposes minimal kinematic restraints to the formation of rupture layers, confirms the existence of the intermediate direction, see [Arthur, Chua and Dunstan, 1977] and [Arthur *et al.*, 1977]. However, such test equipment still needs reinforced corners and — due to the fact that rupture layers tend to originate in the corner — other directions are obtained as well.

There is good evidence that the no-extension-direction is frequently found in situations with one rigid and one open boundary. In this case a stable hyperbolic equation is established: [Roscoe, 1970], [James and Bransby, 1970], in which lines of zero extension characteristics occur prior to rupture layer formation. This work shows that the scenario outlined under point 1, above, is plausible.

Other analyses to find the direction of rupture have been pursued. For example, one might consider under what constitutive circumstances certain jump conditions in stress or strain are permitted. Invariably these analyses do not illuminate the existence of the meta-stable régime. Only an analysis based on quasi-static equilibrium does that.

The fact that all the rupture phenomena take place at, or very near, peak stress makes this analysis insufficiently subtle. Strictly speaking a higher order model should be used, which is also able to give the thickness of the rupture layer (a higher order model includes an extra length scale). The difficulty with higher order models is that they require a lot of parameters, which may be near-impossible to measure.

3.3 Wave speeds and strong ellipticity

The analysis may be extended by considering dynamic processes. Instead of the static equilibrium condition (ignoring Jaumann derivatives for the moment)

$$X_{ijk\ell}\frac{\partial e_{k\ell}}{\partial x_j} = 0$$

The equation of motion for a material with mass density ρ reads

$$X_{ijk\ell}\frac{\partial e_{k\ell}}{\partial x_j} = \rho\frac{\partial^2 u_i}{\partial t^2}$$

This equation has a wave-type solution; the amplitude and polarisation of the wave are described by a vector $\mathbf{A}$, the circular frequency is ω, the wave vector is $\mathbf{k}$ and the phase χ

$$u_i(\mathbf{x},t) = A_i\cos(\omega t + \mathbf{k}.\mathbf{x} + \chi)$$

The wave has a propagating direction $\mathbf{n}$. The wave number k is such that $\mathbf{k} = k\mathbf{n}$; the circular frequency and the wave number are related via the wave speed c_0 as

$$k = \frac{\omega}{c_0}$$

Substituting back into the equation of motion (using the symmetry relations of $\mathbf{X}$) then results in

$$\left(X_{ijpq}n_jn_q - \rho c_0^2\delta_{ip}\right)A_p = 0$$

Now take the inner product with $\mathbf{A}$

$$X_{ijpq}n_jn_qA_iA_p = \rho c_0^2A^2$$

Note that $\mathbf{A}$ must be a real vector, so if the wave speeds are real (meaning that the material can actually transmit a wave) then the following condition holds

$$X_{ijpq}n_jn_qA_iA_p > 0$$

This must be true for all directions and all possible amplitude vectors. The condition is known as the *strong ellipticity condition* and, as its name implies, is somewhat more restrictive than the ellipticity condition encountered before. It imposes limitations on the values that the stiffness components can attain. For a transverse anisotropic medium these are elaborated by [Koenders, 1984]. Static implications of the strong ellipticity condition are treated by [Hayes, 1969]. Related issues of uniqueness and infinitesimal stability are discussed by [Knops and Payne, 1971].

References

Arthur, J.R.F., Dunstan, T., Al-Ani, Q.A.J. and Assadi, A. (1977) Plastic deformation and failure in granular media. *Géotechnique* **27**(1) 53–74.

Arthur, J.R.F., Chua, K.S., and Dunstan, T. (1977) Induced anisotropy in a sand. *Géotechnique* **27**(1) 13–30.

Arthur, J.R.F. and Dunstan, T. (1982) Rupture layers in granular media. *Proc. IUTAM Conference on Deformation and Failure of Granular Materials*, Delft (Vermeer, P.A. Luger, H.J. *eds.*) Rotterdam: Balkema.

Biot, M.A. (1965) *Mechanics of Incremental Deformation*. New York: Wiley.

De Josselin de Jong, G. (1977) Mathematical elaboration of the DSFR model. *Arch. Mech.* **29**(4) 561–591.

Dèsrues, J. and Viggiani, G. (2004) Strain localization in sand: an overview of the experimental results obtained in Grenoble using stereophotogrammetry. *Int. Journal for Numerical and Analytical Methods in Geomechanics* **28**(4) 279–321.

Hayes, M. (1979) Static implications of the strong ellipticity condition. *Archives Rational Mech. Anal.* **33** 181–191.

Hill, R. and Hutchinson, J.W. (1975) Bifurcation phenomena in the plane tension test. *J. Mech. Phys. Solids* **23** 239–264.

Jackson, J.D. (1962) *Classical Electrodynamics*. New York: John Wiley and Sons.

James, R.G. and Bransby, P.L. (1970) Experimental and theoretical investigations of a passive earth pressure problem. *Géotechnique* **20**(1) 17–37.

Knops, R.J. and Payne, L.E. (1971) *Uniqueness Theorems in Linear Elasticity*. Berlin: Springer.

Koenders, M.A. (1984) A two dimensional non-homogeneous deformation model for sand. PhD Thesis, University College London.

Morse, P.M. and Feshbach, H. (1953) *Methods of Theoretical Physics*. New York: McGraw-Hill.

Roscoe, K.H. (1970) The influence of strains in soil mechanics. *Géotechnique* **20**(2) 129–170.

Rudnicki, J.W. and Rice, J.R. (1975) Conditions for the localization of the deformation in pressure-sensitive dilatant materials. *J. Mech. Phys. Solids.* **23** 371–394.

Spencer, A.J.M. (1964) A theory of the kinematics of ideal soils under plane strain conditions. *J. Mech. Phys. Solids* **12** 337–351.

Vardoulakis, I., Goldscheider, M. and Gudehus, G. (1978) Formation of shear bands in sand bodies as a bifurcation problem. *Int. J. Num. Anal. Meth. Geomechanics* **2** 99–128.

Vardoulakis, I. (1979) Bifurcation analysis of the triaxial test on sand samples. *Acta Mechanica* **32** 35–54.

Vardoulakis, I. (1980) Shear band inclination and shear modulus of sand in biaxial tests. *Int. J. Num. Anal. Meth. in Geomechanics* **4** 103–119.

Vardoulakis, I. and Sulem, J. (1995) *Bifurcation Analysis in Geomechanics*. Glasgow, UK: Blackie Academic and Professional, Chapman and Hall.

Vermeer, P.A. (1990) The orientation of shear bands in a biaxial test. *Géotechnique* **40**(2) 223–236.

Chapter 4

Heterogeneity

4.1 General considerations

The study of heterogeneity in continuum mechanics is a well-developed subject. In order to make the insights relevant to densely packed granular materials it is necessary to go into the details of certain aspects, but it is not intended to give a full review here. The question as to why heterogeneity is a necessary element in the understanding of the mechanics of granular media is easily answered. Just consider a medium consisting of one particle surrounded by its next-door neighbours. It was already demonstrated in Chapter 1 that in a granular packing the number of interacting neighbours and the direction of the branch vectors of each particle varies considerably. It is therefore reasonable to expect that the stiffness tensor that is associated with the mini-continuum that represents the stress-strain response on a small scale is also a fluctuating quantity. Leaving aside for the moment the question of how exactly the stiffness is determined on a particle scale, it is surely helpful to determine the effect of a fluctuating stiffness and to build up an intuition for the impact of fluctuations in the system. This analysis will assist in determining the sensitivity of the overall system to the fluctuating content, such as the variability of stress and strain, the effect on overall stiffness components, the relevance of correlation lengths and the formation of correlated structures as the material evolves in a strain path.

Key to the analysis is the use of the quasi-static equilibrium equations for a stress increment. To simplify the approach the Cauchy increment is taken, as the use of the Jaumann increment is really only required close to the formation of rupture layers. So, starting from the equilibrium equations $\partial \sigma_{ij} / \partial x_j = 0$ (not considering body forces) and a local constitutive relation with stiffness tensor $\mathbf{X}$, which is obviously position-dependent, the equation in question is

$$\frac{\partial}{\partial x_j}\left[X_{ijk\ell}(\mathbf{x}) e_{k\ell}(\mathbf{x})\right] = 0$$

No boundary conditions are specified; instead an average strain is imposed. The average strain is the volume average over a volume V, defined as

$$\bar{e}_{k\ell} = \frac{1}{V}\int_V e_{k\ell}(\mathbf{x}) dV$$

The volume is very, very much larger than the size of the constituents.

The stiffness also has a volume average $\bar{\mathbf{X}}$. The fluctuations are then $\mathbf{X}'(\mathbf{x}) = \mathbf{X}(\mathbf{x}) - \bar{\mathbf{X}}$. Substituting these in the equilibrium equations gives

$$\frac{\partial}{\partial x_j}\left\{\left[\bar{X}_{ijk\ell} + X'_{ijk\ell}(\mathbf{x})\right]\left[\bar{e}_{k\ell} + e'_{k\ell}(\mathbf{x})\right]\right\} = 0 \rightarrow$$

$$\bar{X}_{ijk\ell}\frac{\partial e'_{k\ell}(\mathbf{x})}{\partial x_j} + \frac{\partial X'_{ijk\ell}(\mathbf{x})}{\partial x_j}\bar{e}_{k\ell} + \frac{\partial\left[X'_{ijk\ell}(\mathbf{x}) e'_{k\ell}(\mathbf{x})\right]}{\partial x_j} = 0$$

It is seen that the first two terms are 'first order' in the fluctuations, while the third term contains the product of two fluctuations. So, if the fluctuations are small this term may be neglected — an assumption that has to be at least looked at afterwards. This approximation was first introduced by [Kröner, 1967].

The equation

$$\bar{X}_{ijk\ell}\frac{\partial e'_{k\ell}(\mathbf{x})}{\partial x_j} + \frac{\partial X'_{ijk\ell}(\mathbf{x})}{\partial x_j}\bar{e}_{k\ell} = 0$$

is solved by Fourier transformation (see Appendix Section A4). The Fourier wave number vector is denoted by $\mathbf{k}$ and Fourier transforms are denoted by a ^. The strain fluctuation is

$$e'_{k\ell}(\mathbf{x}) = \frac{1}{2}\left[\frac{\partial u'_k(\mathbf{x})}{\partial x_\ell} + \frac{\partial u'_\ell(\mathbf{x})}{\partial x_k}\right]$$

In the Fourier domain the differential equation takes the form

$$-\frac{1}{2}k_j\left(k_\ell \bar{X}_{ijk\ell}\hat{u}_k + k_k \bar{X}_{ijk\ell}\hat{u}_\ell\right) + ik_j \hat{X}'_{ijk\ell}\bar{e}_{k\ell} = 0$$

Making use of the fact that $\bar{X}_{ijk\ell} = \bar{X}_{ij\ell k}$ this may be written as

$$-k_j k_k \bar{X}_{ij\ell k}\hat{u}_\ell + ik_j \hat{X}'_{ijk\ell}\bar{e}_{k\ell} = 0$$

With solution

$$\hat{u}_a = iP_{ai}^{-1}k_j \hat{X}'_{ijk\ell}\bar{e}_{k\ell},$$

where $P_{i\ell} = k_j k_k \bar{X}_{ij\ell k}$, is the 2-tensor known as the *acoustic tensor*, just as in Section 3.1.

No solution exists when $\det(\mathbf{P}) = 0$ and in the previous chapter it was shown that this is exactly the criterion for rupture formation. Another interpretation of the rupture criterion is therefore that the continuum is brought in a state where fluctuations blow up and their effect is felt throughout the whole medium. When the equations remain elliptical the effect of a fluctuation is felt in the vicinity of the fluctuation only. These concepts will be further elucidated.

The Fourier transformed of the strain fluctuation is

$$\frac{i}{2}\left(k_b\hat{u}_a + k_a\hat{u}_b\right) = -\frac{1}{2}\left(P_{ai}^{-1}k_j k_b + P_{bi}^{-1}k_j k_a\right)\hat{X}'_{ijk\ell}\bar{e}_{k\ell}$$

The strain fluctuation in the spatial domain is

$$e'_{ab} = -\frac{1}{2(2\pi)^n}\int d^n k e^{i\mathbf{k}.\mathbf{x}}\left(P_{ai}^{-1}k_j k_b + P_{bi}^{-1}k_j k_a\right)\hat{X}'_{ijk\ell}\bar{e}_{k\ell}$$

The Fourier inverse of the term in brackets is the influence function F_{abij}:

$$F_{abij}(\mathbf{x}) = \frac{1}{(2\pi)^n}\int d^n k e^{i\mathbf{k}.\mathbf{x}}\left(P_{ai}^{-1}k_j k_b + P_{bi}^{-1}k_j k_a\right)$$

Then

$$e'_{ab} = -\frac{1}{2(2\pi)^n}\int d^n k e^{i\mathbf{k}.\mathbf{x}}\int d^n y e^{-i\mathbf{k}.\mathbf{y}}F_{abij}(\mathbf{y})\int d^n z e^{-i\mathbf{k}.\mathbf{z}}X'_{ijk\ell}(\mathbf{z})\overline{e}_{k\ell}$$

Integrate over $\mathbf{k}$ and using the definition of the delta function to integrate over $\mathbf{y}$

$$\begin{aligned} e'_{ab}(\mathbf{x}) &= -\frac{1}{2}\int d^n y F_{abij}(\mathbf{y})\int d^n z X'_{ijk\ell}(\mathbf{z})\delta(\mathbf{x}-\mathbf{y}-\mathbf{z})\overline{e}_{k\ell} \\ &= -\frac{1}{2}\int d^n z F_{abij}(\mathbf{x}-\mathbf{z})X'_{ijk\ell}(\mathbf{z})\overline{e}_{k\ell} \end{aligned}$$

The interpretation of this is that the strain fluctuation in location $\mathbf{x}$ is determined by collecting contributions of the stiffness fluctuations, while the weight of these contributions is dependent on the value of the influence function in positions measured from location $\mathbf{x}$. The question is then: what form does the influence function take. This, of course, depends on the average stiffness moduli. Special cases will be examined to get a feel for the outcome.

There are other ways of determining the strain fluctuations in a heterogeneous medium. A widely researched method is an approach in which the fluctuation is regarded as an inclusion in the mean medium. This method was first introduced by [Eshelby, 1957, 1959]. The problem is choosing the shape of the inclusion, as this choice affects the anisotropy of the problem, see also [Walton, 1977].

4.2 Isotropic materials

In the isotropic case the average moduli take the form

$$\overline{X}_{ijk\ell} = \overline{\lambda}\delta_{ij}\delta_{k\ell} + \overline{\mu}\left(\delta_{ik}\delta_{j\ell} + \delta_{i\ell}\delta_{kj}\right)$$

$$P_{i\ell} = k_j k_k \left[\bar{\lambda}\delta_{ij}\delta_{k\ell} + \bar{\mu}\left(\delta_{ik}\delta_{j\ell} + \delta_{i\ell}\delta_{kj}\right)\right] = \left(\bar{\lambda} + \bar{\mu}\right) k_i k_\ell + \bar{\mu}\delta_{i\ell} k^2$$

Independent of the dimension the inverse is

$$P_{ai}^{-1} = -\frac{\left(\bar{\lambda} + \bar{\mu}\right)}{\mu\left(\bar{\lambda} + 2\bar{\mu}\right) k^4} k_a k_i + \frac{1}{\bar{\mu} k^2}\delta_{ai}$$

And so

$$F_{abij}(\mathbf{x}) = \frac{-1}{(2\pi)^n}\int d^n k e^{i\mathbf{k}.\mathbf{x}} \left(\frac{2\left(\bar{\lambda} + \bar{\mu}\right)}{\bar{\mu}\left(\bar{\lambda} + 2\bar{\mu}\right) k^4} k_a k_i k_j k_b \right)$$
$$+ \frac{1}{(2\pi)^n}\int d^n k e^{i\mathbf{k}.\mathbf{x}} \left[\frac{1}{\bar{\mu} k^2}\left(\delta_{ai} k_j k_b + \delta_{bi} k_j k_a\right)\right]$$

In three dimensions the integral can be done (see Appendix, Section A.6.3).

To get a feel for the functional behaviour of the influence function, the result, as an example F_{1111}, is integrated over x_3 (this makes for an easy two-dimensional visualisation) and then the magnitude is plotted (Fig. 4.1). The outcome is

$$G_{1111} = \left| \int_{-\infty}^{\infty} dx_3 F_{1111} \right| = \left| \frac{4\left(x_1^4 + 6x_1^2 x_2^2 - 3x_2^4\right)}{\left(\bar{\lambda} + 2\bar{\mu}\right) r^6} - \frac{2\left(x_1^4 + 12x_1^2 x_2^2 - 5x_2^4\right)}{\bar{\mu} r^6} \right|$$

In this way no sign is shown, but it is clear from the picture that there is a strong peak near the origin.

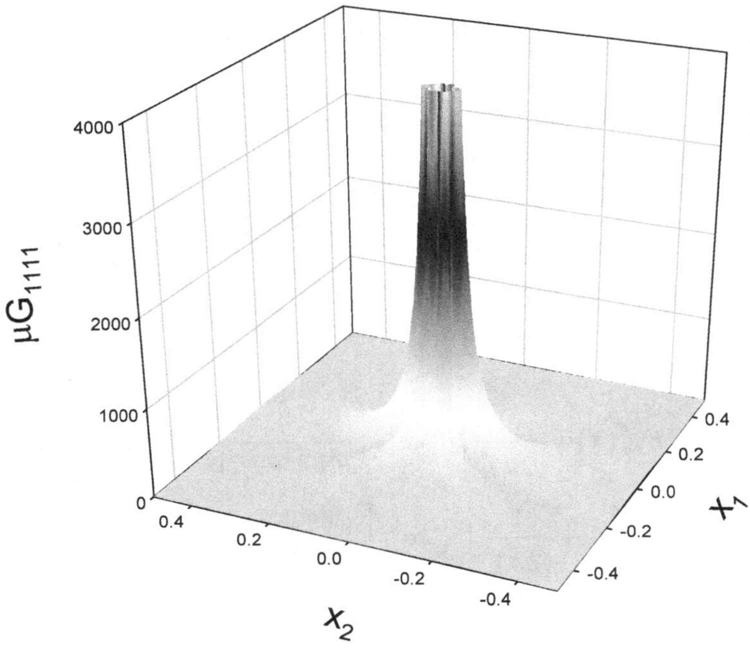

Figure 4.1. Magnitude of the influence function for $\bar{\lambda} = \bar{\mu}/3$.

This behaviour of a strong peak near the origin holds for all components of the influence function, so much so, that it is a good idea to count contributions in the vicinity of the origin as the most dominant ones. It also holds when there is anisotropy — though, naturally, it breaks down when the material exhibits rupture layer formation. In this region around the origin $\mathbf{X}'$ will have a dominant value: $\mathbf{X}'(\mathbf{x})$. Without loss of generality the point $\mathbf{x} = 0$ may be taken (this amounts merely to a translation of the origin). In this approximation the strain fluctuation in the origin becomes

$$e'_{ab}(\mathbf{0}) = -\frac{1}{2}\frac{1}{(2\pi)^n} X'_{ijk\ell}(\mathbf{0}) \int_{\substack{small\\region}} d^n z \int d^n k e^{-i\mathbf{k}.\mathbf{z}} \left(P_{ai}^{-1} k_j k_b + P_{bi}^{-1} k_j k_a\right) \bar{e}_{k\ell},$$

which may be abbreviated as

$$e'_{ab}(\mathbf{0}) = -R_{abij} X'_{ijk\ell}(\mathbf{0}) \overline{e}_{k\ell}$$

The properties of **R**, the *local response tensor*, will be determined more generally below.

To conclude this section it is recalled that the term with double fluctuations has been neglected and the consequences of this approximation should be investigated. The starting point of the analysis was

$$\frac{\partial}{\partial x_j}\left\{\left[\overline{X}_{ijk\ell} + X'_{ijk\ell}(\mathbf{x})\right]\left[\overline{e}_{k\ell} + e'_{k\ell}(\mathbf{x})\right]\right\} = 0 \rightarrow$$

$$\overline{X}_{ijk\ell}\frac{\partial e'_{k\ell}(\mathbf{x})}{\partial x_j} + \frac{\partial X'_{ijk\ell}(\mathbf{x})}{\partial x_j}\overline{e}_{k\ell} + \frac{\partial\left[X'_{ijk\ell}(\mathbf{x})e'_{k\ell}(\mathbf{x})\right]}{\partial x_j} = 0$$

By neglecting the last term the fluctuations $\mathbf{e}'$ are estimated, as demonstrated by the analysis above. Now a fluctuation on top of $\mathbf{e}'$ is considered, which is called $\mathbf{e}''$. Remembering that the equation for $\mathbf{e}'$ is already satisfied. An equation for $\mathbf{e}''$ is obtained

$$\frac{\partial}{\partial x_j}\left\{\left[\overline{X}_{ijk\ell} + X'_{ijk\ell}(\mathbf{x})\right]\left[\overline{e}_{k\ell} + e'_{k\ell}(\mathbf{x}) + e''_{k\ell}(\mathbf{x})\right]\right\} = 0 \rightarrow$$

$$\overline{X}_{ijk\ell}\frac{\partial e'_{k\ell}(\mathbf{x})}{\partial x_j} + \frac{\partial X'_{ijk\ell}(\mathbf{x})}{\partial x_j}\overline{e}_{k\ell} + \overline{X}_{ijk\ell}\frac{\partial e''_{k\ell}(\mathbf{x})}{\partial x_j}$$

$$+\frac{\partial\left[X'_{ijk\ell}(\mathbf{x})e'_{k\ell}(\mathbf{x}) + X'_{ijk\ell}(\mathbf{x})e''_{k\ell}(\mathbf{x})\right]}{\partial x_j} = 0 \rightarrow$$

$$\overline{X}_{ijk\ell}\frac{\partial e''_{k\ell}(\mathbf{x})}{\partial x_j} + \frac{\partial\left[X'_{ijk\ell}(\mathbf{x})e'_{k\ell}(\mathbf{x})\right]}{\partial x_j} + \frac{\partial\left[X'_{ijk\ell}(\mathbf{x})e''_{k\ell}(\mathbf{x})\right]}{\partial x_j} = 0$$

The last term contains the product of higher order fluctuations, compared to the previous terms. As before, this is then neglected and the fascinating result is

$$\overline{X}_{ijk\ell}\frac{\partial e''_{k\ell}(\mathbf{x})}{\partial x_j}+\frac{\partial\left[X'_{ijk\ell}(\mathbf{x})e'_{k\ell}(\mathbf{x})\right]}{\partial x_j}=0$$

This equation has the same form as before, but now the source term $\partial\left(X'_{ijk\ell}(\mathbf{x})\overline{e}_{k\ell}\right)/\partial x_j$ is replaced by $\partial\left[X'_{ijk\ell}(\mathbf{x})e'_{k\ell}(\mathbf{x})\right]/\partial x_j$. The solution to the subsequent approximation is entirely analogous to the one before

$$e''_{ab}(\mathbf{x})=-\frac{1}{2}\int d^n zF_{abij}(\mathbf{x}-\mathbf{z})X'_{ijk\ell}(\mathbf{z})e'_{k\ell}(\mathbf{z})$$

$$=\frac{1}{4}\int d^n zF_{abij}(\mathbf{x}-\mathbf{z})X'_{ijk\ell}(\mathbf{z})\int d^n yF_{k\ell pq}(\mathbf{z}-\mathbf{y})X'_{pqrs}(\mathbf{y})\overline{e}_{rs}$$

It is seen that by this method the higher order strain fluctuation depends quadratically on the stiffness fluctuations. So, while the magnitude of the components of the influence function is less than unity (they will be estimated below) the higher order strain fluctuation is actually quite modest compared to the first order, as long as the magnitude of the stiffness fluctuations is not too large.

In passing it is noted that the volume average of the strain fluctuation is no longer zero when a higher order is considered; $\overline{\mathbf{e}}$ is then a 'first estimate' from which the volume average can be calculated if all the stiffness fluctuations are known.

4.3 Effective stiffness moduli

The mean stress (the volume average) is

$$\overline{\sigma}_{ij}=\overline{X_{ijk\ell}(\mathbf{x})e_{k\ell}}=\overline{X}_{ijk\ell}\overline{e}_{k\ell}+\overline{X'_{ijk\ell}(\mathbf{x})\overline{e}_{k\ell}}+\overline{X}_{ijk\ell}\overline{e'_{k\ell}}+\overline{X'_{ijk\ell}(\mathbf{x})e'_{k\ell}}$$

The volume averages of the single fluctuations are zero and so

$$\overline{\sigma}_{ij}=\overline{X}_{ijk\ell}\overline{e}_{k\ell}+\overline{X'_{ijk\ell}(\mathbf{x})e'_{k\ell}}$$

In the previous section it was seen that the strain fluctuations for small stiffness fluctuations are proportional to the mean strain:

$$e'_{ab}(\mathbf{x}) = -\frac{1}{2}\int d^n z F_{abij}(\mathbf{x}-\mathbf{z}) X'_{ijk\ell}(\mathbf{z})\bar{e}_{k\ell}$$

Therefore, the first non-vanishing term that corrects the volume average of the stiffness is

$$-\frac{1}{2V}\int_V d^n x X'_{ijab}(\mathbf{x})\int d^n z F_{abpq}(\mathbf{x}-\mathbf{z}) X'_{pqk\ell}(\mathbf{z})$$
$$= -\frac{1}{2V}\int_V d^n y F_{abpq}(\mathbf{y})\int d^n z X'_{ijab}(\mathbf{y}+\mathbf{z}) X'_{pqk\ell}(\mathbf{z})$$

The second integral, together with the front factor, is easily recognised as the correlation functions of the stiffness components $S_{ijabpqk\ell}(\mathbf{y})$.

Parenthetically, it is observed that in the Fourier domain the expression takes the short form

$$-\frac{1}{2(2\pi)^{2n}}\int_V d^n y \int d^n t e^{i\mathbf{t}.\mathbf{y}} \hat{F}_{abpq}(\mathbf{t})\int d^n k e^{i\mathbf{k}.\mathbf{y}} \hat{S}_{ijabpqk\ell}(\mathbf{k})$$
$$= -\frac{1}{2(2\pi)^{n}}\int d^n k \hat{F}_{abpq}(\text{-}\mathbf{k}) \hat{S}_{ijabpqk\ell}(\mathbf{k})$$

It was argued in the previous section that the strain fluctuation in a certain location is predominantly determined by the stiffness fluctuation in the immediate vicinity of that location. If that is applied everything simplifies tremendously

$$\bar{\sigma}_{ij} = \bar{X}_{ijk\ell}\bar{e}_{k\ell} - \overline{X'_{ijpq}R_{pqab}X'_{abk\ell}}\bar{e}_{k\ell},$$

where

$$R_{pqab} = \frac{1}{2}\frac{1}{(2\pi)^n}\int_{\substack{small\\region}} d^n z \int d^n k e^{-i\mathbf{k}.\mathbf{z}}\left(P^{-1}_{pa}k_q k_b + P^{-1}_{qb}k_p k_a\right)$$

The integral depends on the *average* stiffness moduli only. The small region may be taken to be a small sphere (3-D) or circle (2-D). In the latter case there are four integrals to be done, two over the spatial coordinates and two over the wave vector components. Note that

$$\int_0^{\rho} dzz \int_0^{2\pi} d\psi e^{-i\mathbf{k.z}} = 2\pi \int_0^{\rho} dzzJ_0(kz) = \frac{2\pi}{k}\rho J_1(k\rho)$$

Then $\mathbf{P}^{-1}$ is proportional to k^{-2}, so the integral over k takes the form

$$2\pi \int_0^{\infty} dkk\frac{1}{k}\rho J_1(k\rho) = 2\pi \int_0^{\infty} d(k\rho)J_1(k\rho) = 2\pi$$

This leaves one integral — the one over the angle in the wave vector space. For the isotropic case this integral is elementary and, collecting the front factors, results in

$$R_{pqab} = \frac{1}{4}\left(\frac{\delta_{qb}\delta_{pa} + \delta_{qa}\delta_{pb}}{\bar{\mu}} - \frac{(\bar{\lambda} + \bar{\mu})(\delta_{qb}\delta_{pa} + \delta_{qa}\delta_{pb} + \delta_{pq}\delta_{ab})}{2(\bar{\lambda} + 2\bar{\mu})\bar{\mu}}\right)$$

In order to illustrate how this plays out, take — for example — the case for which the fluctuations are also isotropic. The effective Lamé coefficients are then

$$\mu = \bar{\mu} - \frac{\overline{(\mu')^2}(\bar{\lambda} + 3\bar{\mu})}{2\bar{\mu}(\bar{\lambda} + 2\bar{\mu})};$$

$$\lambda = \bar{\lambda} - \frac{\overline{(\lambda')^2} + 2\overline{(\lambda'\mu')}(\bar{\lambda} + 3\bar{\mu})}{\bar{\lambda} + 2\bar{\mu}} + \frac{\overline{(\mu')^2}(\bar{\lambda} + \bar{\mu})}{2\bar{\mu}(\bar{\lambda} + 2\bar{\mu})}$$

It is observed that the shear modulus is reduced compared to the volume average. The contraction coefficient may be either reduced or increased.

4.4 Transverse anisotropic material

A refinement of the previous case is to include a measure of transverse anisotropy in the material.

First the inverse of the acoustic tensor is determined

$$\mathbf{P}^{-1} \to \frac{1}{ak_1^4 + bk_1^2k_2^2 + ck_1^4}\begin{pmatrix} \bar{\mu}k_1^2 + \bar{X}_{2222}k_2^2 & -\left(\bar{\mu} + \bar{X}_{1122}\right)k_1k_2 \\ -\left(\bar{\mu} + \bar{X}_{2211}\right)k_1k_2 & \bar{\mu}k_2^2 + \bar{X}_{1111}k_1^2 \end{pmatrix} \equiv \frac{1}{k^2}\widehat{\mathbf{P}}^{-1}$$

$\widehat{\mathbf{P}}^{-1}$ contains all angular information in the wave vector space and a, b and c are defined as before in Section 3.1

$$a = \mu \dot{X}_{2222}\,;\; b = \left(X_{1111}X_{2222} - X_{1122}X_{2211}\right) - \mu\left(X_{2211} + X_{1122}\right);$$

$$c = \mu X_{1111}$$

Now, consider again

$$R_{pqab} = \frac{1}{2}\frac{1}{(2\pi)^n}\int_{\substack{small\\ region}} d^n z \int d^n k e^{-i\mathbf{k}.\mathbf{z}}\left(P_{pa}^{-1}k_q k_b + P_{qb}^{-1}k_p k_a\right)$$

The integrals over the spatial variables and the magnitude of the wave vector k are the same as before, leaving

$$R_{pqab} = -\frac{1}{2}\frac{1}{(2\pi)}\int_0^{2\pi} d\varphi\left(\widehat{P}_{pa}^{-1}\widehat{k}_q\widehat{k}_b + \widehat{P}_{qb}^{-1}\widehat{k}_p\widehat{k}_a\right)$$

Write $c\widehat{k}_1^4 + b\widehat{k}_1^2\widehat{k}_2^2 + a\widehat{k}_2^4 = a\left(y^+\widehat{k}_1^2 - \widehat{k}_2^2\right)\left(y^-\widehat{k}_1^2 - \widehat{k}_2^2\right)$, where $y^\pm = \dfrac{-b \pm \sqrt{b^2 - 4ac}}{2a}$. The integrals are then easily done

$$R_{1111} = \frac{\bar{\mu} - \bar{X}_{2222}}{a - b + c}\,;\; R_{1122} = \frac{\bar{X}_{1122} + \bar{\mu}}{a - b + c}$$

$$R_{1212} = R_{2121} = \frac{1}{2}\left[\frac{\bar{X}_{2222} - \bar{\mu}}{a - b + c} + \frac{\bar{X}_{1111} - \bar{\mu}}{a - b + c}\right]$$

$$R_{1221} = R_{2112} = \frac{1}{2}\left[\frac{\bar{X}_{1122} - \bar{\mu}}{a-b+c} + \frac{\bar{X}_{2211} - \bar{\mu}}{a-b+c}\right]$$

$$R_{1122} = \frac{\bar{X}_{2211} + \bar{\mu}}{a-b+c}; \quad R_{2222} = \frac{\bar{\mu} - \bar{X}_{1111}}{a-b+c}$$

Making

$$a-b+c = -\bar{X}_{1111}\bar{X}_{2222} + \bar{X}_{1122}\bar{X}_{2211} + \bar{\mu}\left(\bar{X}_{1111} + \bar{X}_{2222} + \bar{X}_{1122} + \bar{X}_{2211}\right)$$

The same elements that have been encountered before appear here: the outer determinant and the shear modulus; if these two are small then the local influence tensor will be large. Rupture is in this case understood as the influence of a small heterogeneous element of the continuum propagating through the whole medium.

References

Eshelby, J.D. (1957) The determination of the elastic field of an ellipsoidal inclusion, and related problems. *Proceedings of the Royal Society A* **241**(1226) 376–396.

Eshelby, J.D. (1959) The elastic field outside an ellipsoidal inclusion. *Proceedings of the Royal Society A* **252**(1271) 561–569.

Kröner, E. (1967) Elastic moduli of a perfectly disordered material. *J. Mech Phys. Solids.* **15** 319–329.

Walton, L.J. (1977) The determination of the elastic field of an ellipsoidal inclusion in an anisotropic medium. *Math. Proc. Camb. Phil. Soc.* **81** 283–289.

Chapter 5

Fabric Description

5.1 Voronoi tiling

The primary parameter that describes a granular medium is its *solids volume fraction* ϕ (sometimes — especially in the chemical engineering community — called the *solidosity*). This parameter is defined as the ratio of the solids volume to the total volume. Closely related is the porosity n, the ratio of the pore volume to the total volume: $n = 1 - \phi$. These are of course macroscopic parameters and they say little about the details of the packing. (In passing it is noted that the 'jamming' transition takes place at approximately $\phi \cong 0.84$ for a two-dimensional assembly and $\phi \simeq 0.64$ for a mono-sized three dimensional one.)

An important parameter is the grain size. Mono-sized assemblies are of little practical interest and the *grain size distribution* needs to be specified. In the civil engineering literature — see for example [Terzaghi, Peck and Mesri, 1996] — the distribution is specified by *weight*. In the physics literature a specification by *number* is frequently encountered. In the practice of soil mechanics the grain size distribution is determined by sieving, giving the cumulative distribution. Characterisation of the grain size distribution is often done by a few characteristic numbers: the d_{10} is that size (diameter) below which 10% of the weight of the sample is measured. The d_{15}, d_{60} and d_{85} are similarly defined as the sizes below which 15%, 60% and 85% respectively of the weight of the sample have been determined. The ratio d_{60} / d_{10} is called the *uniformity coefficient*. The translation of a distribution by weight to a distribution by number involves the cube of the diameter, so a substantial skewing of the curve may be expected. For most natural samples the d_{15} corresponds roughly to the mean size by number.

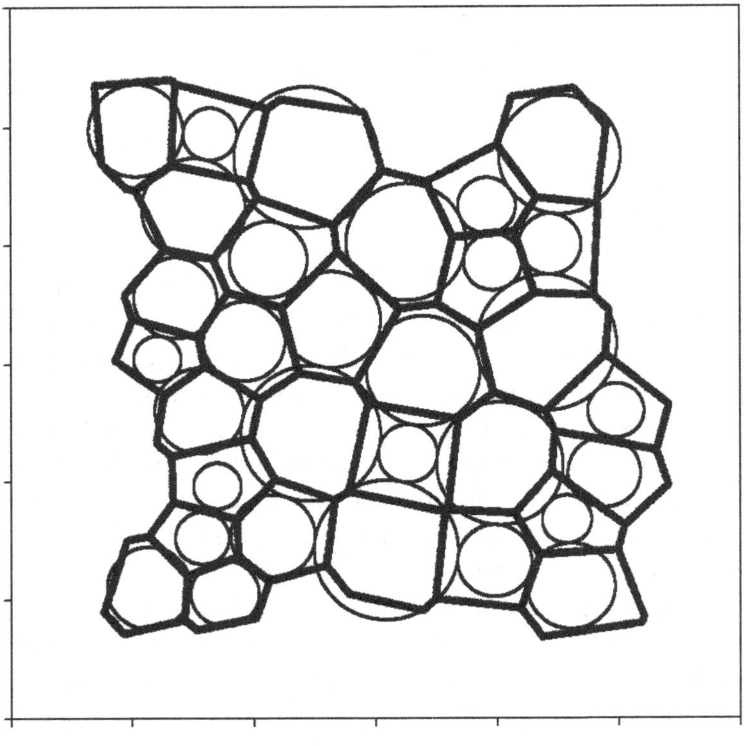

Figure 5.1. Illustration of Voronoi tiling.

A tiling is introduced via the definition of the *Voronoi boundaries* — [Voronoi, 1908]. This definition (which can be used in a far wider context than just for granular materials) defines a boundary as the collection of points that have the shortest distance to a set of given points (called the *generators*). An illustration shown in Fig. 5.1 may be helpful to understand the definition. Take an assembly of discs and for the 'set of given points' use the centres of the discs. The Voronoi boundaries then look as shown in Fig. 5.2. It produces a space-filling tiling. This would be satisfactory for the purposes of granular media if all the discs were of equal size. But because they are not of an equal size the boundaries end up cutting through the particles. The desire is to employ the Voronoi boundaries later on and associate them with particle interactions. The definition of a Voronoi boundary is then changed slightly to ensure that each particle lies within a tile.

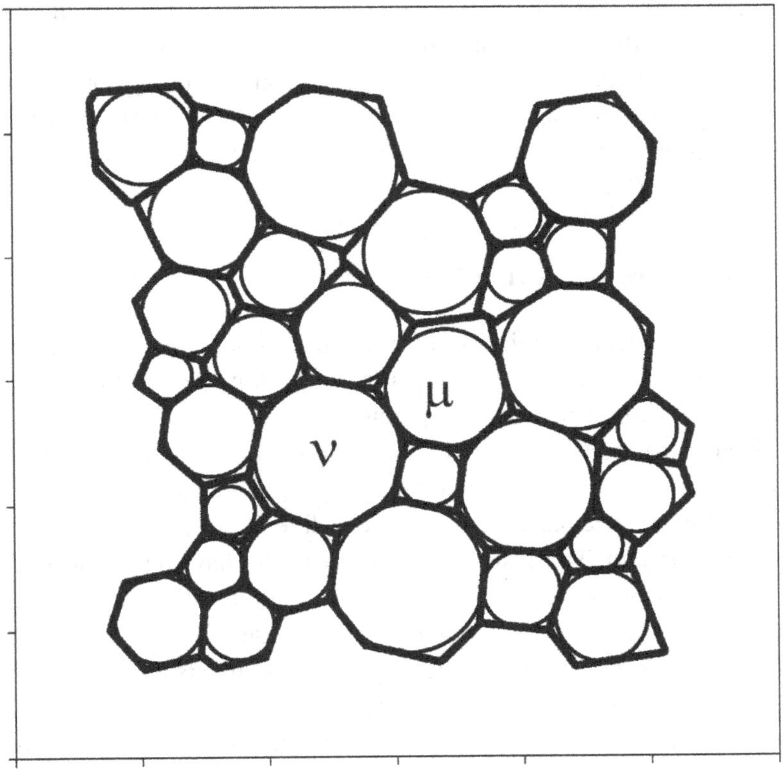

Figure 5.2. Illustration of modified Voronoi tiling ensuring that all boundaries go through the interaction points.

In the modified definition the Voronoi boundaries are drawn in such a way that each point inside the tile is at the shortest distance from a generator greater than, or equal to, the radius associated with that generator (see Fig. 5.2). For slightly overlapping particles the definition needs to be refined again to be sure that the boundary goes through the midpoint of the overlap region. The resulting Voronoi tiling is shown in the figure below. This is a very useful tiling in that each boundary corresponds to an interaction (which may be zero when particles do not touch). The disadvantage of this tiling is that the regions may vary in size quite considerably, whereas for the tiling that follows from the strict mathematical definition above leads to tiles that all have more or less the same area. All particle pairs that share a boundary are *(near) neighbours.*

There are other partitioning methods. The best-known is the *Delaunay triangulation,* which is obtained from the Voronoi tiling by connecting the centres of the near neighbours, see [Delaunay, 1932]. This triangulation contains no information that is not already in the Voronoi tiling: [Liebling and Pournin, 2012].

5.2 Contact point distribution

For the mechanical behaviour of an assembly the location of contact points is important. The tessellations say next to nothing about this aspect and therefore it needs to be treated separately.

When particles are in contact they can transmit a force. The direction of the contacts is important, both on a particle level and — statistically — assembly-wide. The particles have a very diverse set of contacts when the packing is random. To illustrate this, two particles have been selected from the small assembly pictured above. The particle labelled μ has three contacts (another Voronoi boundary comes close, but the two neighbours are just a whisker away from touching). These contacts are depicted in the angular diagram as the straight lines that go to a value of unity. The particle labelled ν has five contacts and these are shown in the second angular diagram. It is seen that the two are very different.

It is possible to characterise the contact distribution by fitting the contacts to a function. This function equals unity when there is a contact in a certain direction and zero when there is no contact. The contacts are then associated with very sharply-peaked functions, which have an area equal to unity; these are delta functions. They are fitted to a particle-specific fabric function of the form (for particle μ) $p^{\mu}(\varphi) = q_i^{\mu} n_i(\varphi) + p_{ij}^{\mu} n_i(\varphi) n_j(\varphi)$ (for particle numbers a Greek superscript is used). This is a reasonable thing to do. In two dimensions the number of coefficients is five. The number of Voronoi boundaries for this particle is six. Each Voronoi boundary is a potential contact and all boundaries enclose the particle. The vector $\mathbf{q}^{\bullet}$ informs on how asymmetric the contacts are arranged, while the tensor $\mathbf{p}^{\bullet}$ gives the total number of contacts and indicates how anisotropically the contacts are distributed.

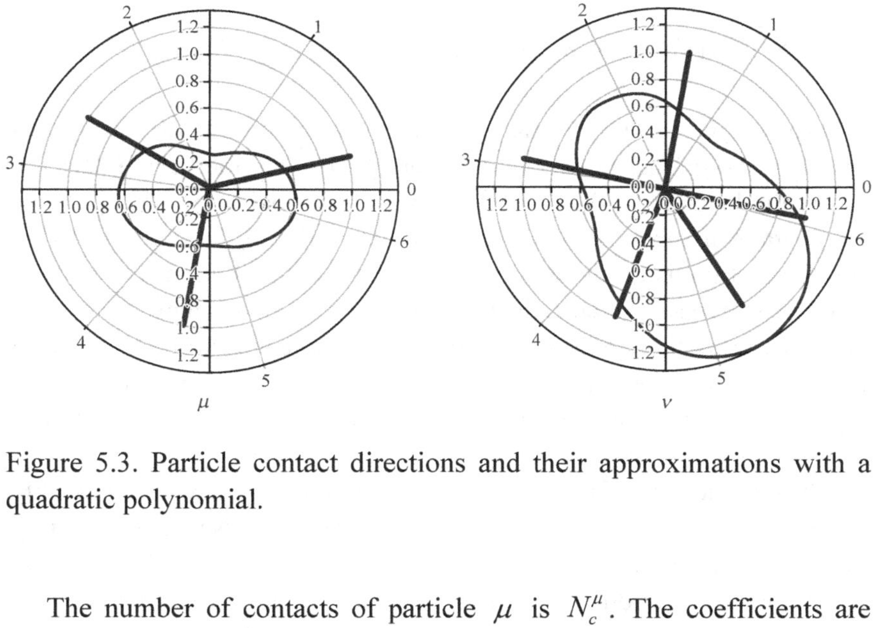

Figure 5.3. Particle contact directions and their approximations with a quadratic polynomial.

The number of contacts of particle μ is N_c^{μ}. The coefficients are obtained from a least-squares fit

$$\int_0^{2\pi}\left[q_i^{\mu}n_i(\varphi)+p_{ij}^{\mu}n_i(\varphi)n_j(\varphi)-\sum_{\varepsilon=1}^{N_c^{\mu}}\delta\left(\varphi-\varphi^{\varepsilon}\right)\right]^2 d\varphi=\min$$

Differentiating with respect to the components of $\mathbf{p}$ and $\mathbf{q}$ yields the set of equations

$$\int_0^{2\pi}\left[q_i^\mu n_i(\varphi)+p_{ij}^\mu n_i(\varphi)n_j(\varphi)-\sum_{\varepsilon=1}^{N_c^\mu}\delta\left(\varphi-\varphi^\varepsilon\right)\right]n_a(\varphi)d\varphi=0\rightarrow$$

$$\pi q_i^\mu\delta_{ia}-\sum_{\varepsilon=1}^{N_c^\mu}n_a\left(\varphi^\varepsilon\right)=0$$

$$\int_0^{2\pi}\left[q_i^\mu n_i(\varphi)+p_{ij}^\mu n_i(\varphi)n_j(\varphi)-\sum_{\varepsilon=1}^{N_c^\mu}\delta\left(\varphi-\varphi^\varepsilon\right)\right]n_a(\varphi)n_b(\varphi)d\varphi=0$$

$$\rightarrow\frac{\pi}{4}p_{ij}^\mu\left(\delta_{ia}\delta_{jb}+\delta_{ib}\delta_{ja}+\delta_{ij}\delta_{ab}\right)-\sum_{\varepsilon=1}^{N_c^\mu}n_a\left(\varphi^\varepsilon\right)n_b\left(\varphi^\varepsilon\right)=0$$

The function $p^\mu(\varphi)$ is also plotted in the polar plots of Fig. 5.3. At first sight the fit does not seem to be very good. The fitted curve does not follow the location of the contacts very well. It does however reproduce the average number of contacts accurately — πp_{ii} — and gives an impression of anisotropy and asymmetry of the distribution. For particle μ there is little asymmetry: the three contacts are quite evenly distributed over the angles, though there is a clearly detectable anisotropy in the East-West direction. For particle ν, on the other hand, the contacts are somewhat bunched towards the South-East and the anisotropy direction is also visible.

The particle fabric functions may be averaged over the whole assembly to give the assembly fabric function. The latter can also be obtained by making a histogram of the contacts, which is frequently done in interpreting simulation results. For a statistically uniform assembly, the asymmetry coefficients tend to average to zero. Thus information is suppressed that could be useful. The squares of the asymmetry components could play a role in understanding the behaviour of the assembly.

The examples given here are all in two dimensions. However, the whole analysis is just as easily pursued in three dimensions (though this is more difficult to visualise). The integrals required to evaluate the least squares minimisation in 3-D are

$$\int n_i n_j d\Omega = \frac{4\pi}{3}\delta_{ij} \text{ and } \int n_i n_j n_k n_\ell d\Omega = \frac{4\pi}{15}\left(\delta_{ij}\delta_{k\ell} + \delta_{ik}\delta_{j\ell} + \delta_{i\ell}\delta_{jk}\right)$$

The term *fabric tensor* has been introduced in the literature and can be obtained from the contact distribution tensor by calculating $\mathbf{p}/p_{ii}$, see [Satake, 1982] for an early reference.

5.3 Correlation

One of the intriguing aspects of granular materials in a random packing is whether there are correlations in the arrangement of the grains. The question really is whether random, densely packed granular materials try to approximate a crystal structure. Is there a tendency for contacts to line up in a direction, possibly in a small environment, comprising a handful of particles? In order to ascertain that aspect, a packing is subjected to an analysis of a correlation function. It has already been seen that two Voronoi boundaries can come close, but not touch. In order to answer the questions about ordering of some sort, a correlation function based on contacts is therefore not adequate. A correlation function based on the distance from the centre of a particle to the Voronoi boundary in a given direction is more suitable.

The distance from a particle centre to the nearest Voronoi boundary is called d_V; a superscript is added to identify the particle. The correlation function that is evaluated is

$$\tilde{\phi}_{d_V}(\varphi) = \frac{1}{N}\sum_{\mu=1}^{N}\int_0^{2\pi} d_V^\mu(\vartheta+\varphi)\, d_V^\mu(\vartheta)\, d\vartheta \text{ normalised to}$$

$$\phi_{d_V}(\varphi) = \tilde{\phi}_{d_V}(\varphi)/\tilde{\phi}_{d_V}(0)$$

The calculation of the correlation function is done for some 4000 particles in two dimensions of a compressed sample and then plotted, as shown in Fig. 5.4. The result is interesting. The correlation function is symmetric: $\phi_{d_V}(-\varphi) = \phi_{d_V}(\varphi)$, so the interval $0 \le \varphi < \pi$ needs to be studied only. As the angle increases a negative correlation is observed first. That is as expected: if two Voronoi boundaries are very close together then the angles directly next to them must be further away. That is a direct consequence of the convexity of the grains. After the initial negative correlation, when the angle has reached about $\pi/2$, all correlation appears to be lost, other than some slight negative noise. The important conclusion is that positional correlations persist as far as near-neighbours, but no further. A hetero-disperse granular assembly has no inherent crystal structure. This does not mean to say that the contacts or particle interactions cannot form a structure. The aligned force chains that emerge when a granular medium is deviatorically stressed are an example of this. Generally, any random medium that is subject to a deviatoric load and evolves locally according to the strain must form structures of some sort, see [Koenders, 1997]. Also, the rupture layer that forms at or near peak stress, points to collective behaviour of the motion of the particles, which is strongly correlated.

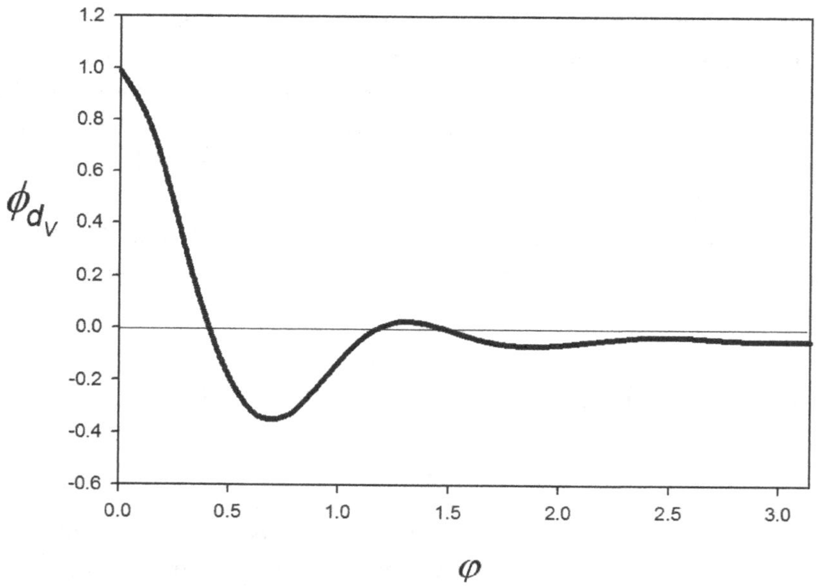

Figure 5.4. Normalised correlation function of the angular Voronoi boundary distribution.

5.4 Strain in a granular medium

The continuum definition of the deformation gradient can be extended to a granular medium by letting it describe the motion of the particle centres (chosen, for example, as the centre of gravity of the grains). The simplest manner in which this can be done, introduced by [Koenders, 1994], is by fitting the displacement $\mathbf{u}^{\mu}$ on the surface of a group of particles to the average deformation gradient $\boldsymbol{\alpha}$ by requiring

$$\sum_{\mu}\left(u_i^{\mu}-u_i(0)-\alpha_{ij}x_j^{\mu}\right)^2 \rightarrow \text{minimal}$$

Setting the derivatives with respect to the components of $\boldsymbol{\alpha}$ to zero leads to the result

$$\alpha_{ij}=\left(\sum_{\mu}\mathbf{x}^{\mu}\mathbf{x}^{\mu}\right)_{jk}^{-1}\sum_{\mu}\left[u_i^{\mu}-u_i(0)\right]x_k^{\mu}$$

This does not look like the continuum definition at all. To make a connection between the discrete definition introduced here and the continuum definition, the surface is supposed to be populated with particles with an angular density $\ell(\Omega)$, this could be either an angle in 2-D (number per radian), or a solid angle in 3-D (number per solid angle area). For a large enough assembly this may be replaced by the average ℓ. The summations may then be replaced by integrals

$$\sum_{\mu}x_i^{\mu}x_j^{\mu} \rightarrow \oint_{sphere/disc} \ell R^2 R^{d-2} n_i n_j d\Omega = \ell V \delta_{ij}$$

$$\sum_{\mu}\left[u_i^{\mu}-u_i(0)\right]x_k^{\mu} \rightarrow \oint_{\substack{unit\\ sphere/disc}} \left[u_i(\Omega)-u_i(0)\right]\ell R R^{d-2} n_k d\Omega$$

The term $RR^{d-2}d\Omega$ is just an infinitesimal area (3-D) or line element (2-D). Using Stokes's theorem the integral becomes

$$\oint_{\substack{unit \\ sphere/disc}} \left[u_i(\Omega) - u_i(0)\right] \ell R R^{d-2} n_k d\Omega =$$

$$\ell \oint_{sphere/disc} \left[u_i(\Omega) - u_i(0)\right] n_k dA = \ell \oint_{sphere/disc} \frac{\partial u_i}{\partial x_k} dV$$

The discrete definition is therefore equivalent to

$$\alpha_{ik} = \frac{1}{V} \oint_{sphere/disc} \frac{\partial u_i}{\partial x_k} dV$$

This is as expected from the volume average of the continuum definition. However, the discrete definition can be employed on *any* assembly of particles; it does not have to be a large assembly; it could be applied to a single Voronoi cell. At the same time the deformation gradient in the continuum interpretation is the first term in a Taylor series of the displacement. This interpretation then remains intact when the discrete definition is used for a small assembly.

The strain is the symmetric part of $\boldsymbol{\alpha}$.

5.5 Stress in a granular medium

The stress may also be defined from a continuum definition, while the medium is in equilibrium. Various contributors have studied this problem: [Love, 1934], [Dantu, 1968] and [Bagi, 1996], while the approach taken below closely follows [Drescher and De Josselin de Jong, 1972]. The particles are stressed by contact forces. There is then a — very complicated — stress field in the system of grains and pores. The stress in the empty pores is zero. The overall stress is

$$\bar{\Sigma}_{ij} = \frac{1}{V} \oint_{volume} \Sigma_{ij}(\mathbf{x}) dV$$

The volume may be partitioned in Voronoi cells

$$\bar{\Sigma}_{ij} = \frac{1}{V}\sum_{\mu} \oint_{volume\, V^{\mu}} \Sigma_{ij}(\mathbf{x})\,dV$$

Now, the equilibrium equations for static equilibrium are known $\partial \Sigma_{ij} / \partial x_j = 0$; using this the stress can be written as a gradient: $\partial\left(x_j \Sigma_{ik}\right) / \partial x_k = 0$. Stokes' theorem is again applied to give

$$\bar{\Sigma}_{ij} = \frac{1}{V}\sum_{\mu} \oint_{volume\, V^{\mu}} \Sigma_{ij}(\mathbf{x})\,dV = \frac{1}{V}\sum_{\mu} \oint_{volume\, V^{\mu}} \frac{\partial\left(\Sigma_{ik}(\mathbf{x})\, x_j\right)}{\partial x_k}\,dV$$

$$= \frac{1}{V}\sum_{\mu} \oint_{area\, A^{\mu}} \Sigma_{ik}(\mathbf{x})\, x_j n_k\, dA$$

$\Sigma_{ik}(\mathbf{x}) n_k$ is precisely the definition of the traction. The Voronoi cell walls go through the contact points and the integral of the traction over the (very small) contact area of each contact point is the contact force. The integral therefore becomes a sum over contacts

$$\bar{\Sigma}_{ij} = \frac{1}{V}\sum_{\mu} \oint_{volume\, V^{\mu}} \Sigma_{ij}(\mathbf{x})\,dV = \frac{1}{V}\sum_{\mu} \oint_{volume\, V^{\mu}} \frac{\partial\left(\Sigma_{ik}(\mathbf{x})\, x_j\right)}{\partial x_k}\,dV$$

$$= \frac{1}{V}\sum_{\mu}\sum_{\nu=1}^{N^{\mu}} F_i^{\mu\nu} x_j^{\mu\nu},$$

where the notation $\mathbf{x}^{\mu\nu} = \frac{1}{2}\left(\mathbf{x}^{\nu} - \mathbf{x}^{\mu}\right)$ has been introduced.

Various manipulations can be done. Using the fact that each particle must be in equilibrium $\sum_{\nu=1}^{N^{\mu}} F_i^{\mu\nu} = 0$, a constant vector (that is, independent of particle number ν) $\mathbf{y}^{\mu}$ may be added to the vector $\mathbf{x}^{\mu\nu}$. The origin of the coordinate frame inside each Voronoi cell is therefore arbitrary and the choice is free (the centre of gravity of the particle, the centre of the Voronoi cell, *etc*). The double sum can also be cast in the form of a sum over contacts. Each force (μ, ν) is encountered twice, once with the location vector $\mathbf{x}^{\mu\nu}$ and once with location vector $\mathbf{x}^{\nu\mu}$. Remembering that $\mathbf{F}^{\mu\nu} = -\mathbf{F}^{\nu\mu}$ and that $\mathbf{x}^{\mu\nu} - \mathbf{x}^{\nu\mu} = \mathbf{c}^{\mu\nu}$, the branch vector, it follows that

$$\frac{1}{V}\sum_{\mu}\sum_{\nu=1}^{N^{\mu}} F_i^{\mu\nu} x_j^{\mu\nu} = \frac{1}{V}\sum_{pairs\,(\mu,\nu)} F_i^{\mu\nu} c_j^{\mu\nu}$$

This can be recast again in the form of a sum over the individual particles, using $\mathbf{F}^{\mu\nu} = -\mathbf{F}^{\nu\mu}$ and $\mathbf{c}^{\mu\nu} = -\mathbf{c}^{\nu\mu}$

$$\bar{\Sigma}_{ij} = \frac{1}{2V}\sum_{\mu}\sum_{\nu=1}^{N^{\mu}} F_i^{\mu\nu} c_j^{\mu\nu}$$

For each particle moment equilibrium is expressed as

$$\varepsilon_{kij}\sum_{\nu=1}^{N^{\mu}} F_i^{\mu\nu} x_j^{\mu\nu} = 0$$

Consequently

$$\sum_{\nu=1}^{N^{\mu}} F_i^{\mu\nu} x_j^{\mu\nu} - \sum_{\nu=1}^{N^{\mu}} F_j^{\mu\nu} x_i^{\mu\nu} = 0$$

The stress tensor $\mathbf{\Sigma}$ is therefore guaranteed symmetric. The basis for this is the same as the one encountered in continuum mechanics.

5.6 Calculating averages in a contacting granular material

Bearing in mind that the formula for the stress only contains as many equations as there are stress components (four in 2-D and nine in 3-D), a first approximation of the contact forces from a given overall stress state can only be quite primitive. A tensor $\hat{\mathbf{T}}$ is introduced ($\hat{\mathbf{t}}$ for increments) in such a way that a contact force is given by an average plus corrections

$$F_i^{\mu\nu} = \hat{T}_{ij} n_j^{\mu\nu} + F_i'^{\mu\nu}$$

The tensor $\hat{\mathbf{T}}$ only describes the force across a *contacting* particle pair, as opposed to the stress tensor, which describes the force across a Voronoi boundary. The latter is zero in the absence of a touching contact.

The contact point distribution tensor $p(\varphi)$ can be viewed as a mapping from the Voronoi boundary distribution to the contact distribution.

Substituting this back into the assembly stress formula the following is obtained

$$\bar{\Sigma}_{ij} = \frac{1}{V}\hat{T}_{ik}\sum_{\mu}\sum_{\substack{\nu=1\\(contacts)}}^{N^\mu} n_k^{\mu\nu}x_j^{\mu\nu} + \frac{1}{V}\sum_{\mu}\sum_{\substack{\nu=1\\(contacts)}}^{N^\mu} F_i'^{\mu\nu}x_j^{\mu\nu}$$

Introducing then a 'cell contact radius' that approximates $\mathbf{x}^{\mu\nu} = \hat{R}^\mu \mathbf{n}^{\mu\nu}$, permits the evaluation of the inner sum using the contact distribution tensor $\mathbf{p}^\mu$, the mean stress is

$$\bar{\Sigma}_{ij} = \frac{\aleph^{(4,d)}}{V}\hat{T}_{ik}\sum_{\mu}\hat{R}^\mu p_{\ell m}^\mu\left(\delta_{\ell m}\delta_{kj} + \delta_{\ell k}\delta_{mj} + \delta_{\ell j}\delta_{km}\right) + \frac{1}{V}\sum_{\mu}\sum_{\nu=1}^{N^\mu} F_i'^{\mu\nu}x_j^{\mu\nu},$$

where $\aleph^{(4,2)} = \frac{\pi}{4}$ in 2-d and $\aleph^{(4,3)} = \frac{4\pi}{15}$ in 3-d, see Appendix, Section A2.

If in first approximation the term that contains $\mathbf{F}'$ is neglected, the tensor $\hat{\mathbf{T}}$ can be found from averaging and inverting,

$$\bar{\Sigma}_{ij} \simeq \frac{\aleph^{(4,d)}}{V}\hat{T}_{ik}\sum_{\mu}\hat{R}^\mu\left(p_{mm}^\mu\delta_{kj} + 2p_{kj}^\mu\right) = \frac{\aleph^{(4,d)}}{V}\hat{T}_{ik}N\left(\overline{\hat{R}p_{mm}}\delta_{kj} + 2\overline{\hat{R}p_{jk}}\right)$$

$$\hat{T}_{i\ell} \simeq \frac{V}{N\aleph^{(4,d)}}\bar{\Sigma}_{ij}\left(\overline{\hat{R}p_{mm}}\boldsymbol{\delta} + 2\overline{\hat{R}\mathbf{p}}\right)_{j\ell}^{-1} \tag{5.1}$$

If this estimate is taken as adequate, then it follows that for an anisotropic contact distribution the assembly stress ratio is larger than the force ratio at the particle scale.

All the above for the stress and the deformation gradient remains true when increments are considered. Simply replace the capital symbols with small-type symbols.

The averaging in this case is over *contacting* particles and when considering a sum over contacts it is permitted to replace the sum by an integral, weighed with the contact distribution. For a contact quantity q, summed over the contacts of a particle the replacement takes the form

$$\sum_{\nu=1}^{N^\mu} q^{\mu\nu} \rightarrow \int_{\textit{circle or sphere}} q(\Omega)p^\mu(\Omega)d\Omega$$

For an individual particle, noting that the contact point description is so poor, this procedure does not make a lot of sense. However, for larger-scale averages it is perfectly acceptable.

Other averages that will be encountered are averages over the boundaries of the Voronoi cell. The quantities in question will generally be geometrical in nature. The question is then how an appropriate average will be defined. Consider particle μ in Fig. 5.2. It has a very short boundary to the North-East, which is quite far from the centre of the particle compared to the other boundaries. Surely, the contribution to a sum over the boundaries of this particular one should carry less weight than the other ones. Calling the length of the boundaries $\ell^{\mu\bullet}$, the average of a quantity q, which is defined on the boundaries, would be weighed with the length of the boundary. In other words

$$\overline{q}^\mu = \frac{1}{N_V^\mu}\sum_{\varepsilon=1}^{N_V^\mu} q^{\mu\varepsilon} \rightarrow \frac{1}{N_V^\mu \overline{\ell}^\mu}\sum_{\varepsilon=1}^{N_V^\mu} q^{\mu\varepsilon}\ell^{\mu\varepsilon},$$

where $\overline{\ell}^\mu$ is the mean length of the Voronoi boundaries.

A case in point is $q^{\mu\varepsilon} \rightarrow \mathbf{c}^{\mu\varepsilon}\mathbf{c}^{\mu\varepsilon}$, the two-tensor of the branch vectors. The angular part is quite easily dealt with, assuming that the angles are more or less isotropically positioned, that is they are distributed according to a distribution with angular distribution $N_V^\mu / 2\pi$. Then

$$\frac{1}{N_V^\mu \overline{\ell}^\mu}\sum_{\varepsilon=1}^{N_V^\mu} c_i^{\mu\varepsilon} c_j^{\mu\varepsilon}\ell^{\mu\varepsilon} = \frac{1}{N_V^\mu \overline{\ell}^\mu}\sum_{\varepsilon=1}^{N_V^\mu} \left(c^{\mu\varepsilon}\right)^2 n_i^{\mu\varepsilon} n_j^{\mu\varepsilon}\ell^{\mu\varepsilon} \rightarrow$$

$$\frac{1}{N_V^\mu \overline{\ell}^\mu}\sum_{\varepsilon=1}^{N_V^\mu} \left(c^{\mu\varepsilon}\right)^2 n_i^{\mu\varepsilon} n_j^{\mu\varepsilon}\ell^{\mu\varepsilon} \rightarrow \frac{1}{N_V^\mu}\frac{N_V^\mu}{2\pi}\int_0^{2\pi} n_i(\varphi)n_j(\varphi)d\varphi \overline{\left(c^{\mu\bullet}\right)^2 \frac{\ell^{\mu\bullet}}{\overline{\ell}^\mu}}^\mu =$$

$$= \frac{\delta_{ij}}{2}\overline{\left(c^{\mu\bullet}\right)^2 \frac{\ell^{\mu\bullet}}{\overline{\ell}^\mu}}^\mu$$

The term $\overline{\left(c^{\mu\bullet}\right)^2 \frac{\ell^{\mu\bullet}}{\overline{\ell^{\mu}}}}^{\mu}$ is the weighted average of the quadratic lengths of the branch vectors. It represents another quadratic length scale: $\overline{c^2}^{\mu}$.

All this is in two dimension, however, the analysis goes in exactly the same way in three dimensions. Replace the length of the Voronoi boundary by its area and use the appropriate front factor $\aleph^{(2,3)}$, rather than $\aleph^{(2,2)}$.

References

Bagi, K. (1996) Stress and strain in granular assemblies. *Mechanics of Materials* **22** 165–177.

Dantu, P. (1968) Étude statistique des forces intergranulaires dans un milieu pulvérument. *Géotechnique* **18** 50–55.

Delaunay, B.N. (1932) Neue Darstellung der geometrischen Kristallographie. *Zeitschrift für Kristallograph* **84** 109–149.

Drescher, A. and De Josselin de Jong, G. (1972) Photoelastic verification of a mechanical model for the flow of a granular material. *Journal of the Mechanics and Physics of Solids* **20**(5) 337–340.

Koenders, M.A. (1994) Least squares methods for the mechanics of nonhomogeneous granular assemblies. *Acta Mechanica* **106** 23–40.

Koenders, M.A. (1997) The evolution of spatially structured elastic materials using a harmonic density function. *Phys Rev E* **56**(5) 5585–5593.

Liebling, T.M. and Pournin, L. (2012) Voronoi diagrams and Delaunay triangulations: Ubiquitous siamese twins. *Documenta Mathematica*, **Extra Volume ISMP** 419–431.

Love, A.E.H. (1934) *A Treatise on the Mathematical Theory of Elasticity*. Cambridge: CUP.

Satake, M. (1982) Fabric tensor in granular materials. *Proc. IUTAM Conference on Deformation and Failure of Granular Materials*, Delft (Vermeer, P.A. and Luger, H.J. *eds.*) Rotterdam: Balkema.

Terzaghi, K., Peck, R.B. and Mesri, G. (1996) *Soil Mechanics in Engineering Practice*. New York: John Wiley & Sons.

Voronoi, G. (1908) Nouvelles applications des paramètres continus à la théorie des formes quadratiques. *J. Reine Angew. Math.* **134** 198–287.

Chapter 6

Stress-Strain Relations of Granular Assemblies: A Frictionless Assembly

6.1 General considerations

The purpose of this and the following chapters is to make a link between the micro-mechanical details and the overall, assembly-averaged properties. The latter consists of an estimate of the stiffness tensor that connects incremental strain and incremental stress. The case of an assembly in which the particles interact solely through normal movement is considerably simpler than the more general case in which a tangential interaction also needs to be considered. The theory for frictionless assemblies is less involved — and therefore much more transparent — than the one in which a tangential interactive component needs to be accounted for as well, as the particle spins are not restricted and no rotational equilibrium equations are needed. This provides an opportunity to develop the main theoretical concepts, which can later be used to describe more involved interactive properties.

The case of frictionless particles is not merely a dry mathematical exercise; it has a number of practical applications as well. For assemblies in which there are more contacts per particle than the isostatic requirement, an incremental stiffness exists. This case is relevant to small spherical particles in a fluid environment, the interaction for which was discussed in Chapter 1, Section 1.7. In the field of chemical engineering, where this subject has major relevance, the theory has been employed to describe the behaviour of cakes composed of micron-sized particles. Such cakes appear, for example, in filtration processes and in slurries, see [Wakeman and Tarleton, 1999], [Koenders and Wakeman, 1997] and [Civan, 2007].

The theory presented here was initially developed by [Koenders, 1984, 1987] and further refined by [Jenkins and Koenders, 2004]. Essentially, the question is what the influence of heterogeneity is on the assembly-averaged (incremental) stiffness tensor. To answer this question a mean-field estimate of the stiffness is calculated first and then a subsequent correction due to fluctuations is introduced. The procedure is similar to, but subtly different from, the approach put forward in Chapter 4 for continua. The influence of the heterogeneity is captured in a differential equation, the solution of which yields a correction to the displacement in addition to the mean strain displacement. The correction to the displacement gives a force correction, which feeds into a stress correction. *All quantities are incremental*, because of the expected severe non-linearity of the contact law.

6.2 Kinematics

The displacement in the vicinity of particle μ can be expanded in a Taylor series in the branch vectors $\mathbf{c}^{\mu\bullet}$ (defined in Section 5.5). The branch vector is related to the particle positions as $\mathbf{c}^{\mu\nu} = \mathbf{x}^{\mu\nu} - \mathbf{x}^{\nu\mu} = \mathbf{x}^{\nu} - \mathbf{x}^{\mu}$. The displacement increment of a neighbouring particle can be approximated as

$$u_i^{\nu} \simeq u_i^{\mu} + \left.\frac{\partial u_i}{\partial x_j}\right|^{\mu} c_j^{\mu\nu} + \frac{1}{2}\left.\frac{\partial^2 u_i}{\partial x_j \partial x_k}\right|^{\mu} c_j^{\mu\nu} c_k^{\mu\nu}$$

It is sensible to ascertain how many terms need to be taken account of. To that end it is established how many parameters are implied in the Taylor series. For constitutive purposes the displacement *difference* is relevant, that is $\mathbf{u}^{\nu} - \mathbf{u}^{\mu}$, so the Taylor series is essentially expanding this quantity. The purpose is to describe the motion of particles in the vicinity of particle μ and therefore those neighbours that share a Voronoi boundary should be represented. Table 6.1 gathers the number of independent displacement components for each term in the Taylor series.

Table 6.1. Number of displacement components for each term in the Taylor series.

	First derivative	Second derivative	Total number
2 dimensions	2	3	5
3 dimensions	3	6	9

Taking the series up to the second derivative permits the specification of five particles in 2-D and nine particles in 3-D. Fewer terms in the Taylor series would under-specify the motion of the neighbours, but taking account of higher orders really gives rise to over-specification.

The second derivative can be obtained from a least squares fit of neighbouring particles and expressed in terms of the first derivatives by requiring

$$\sum_{\nu=1}^{N^\mu}\left(\left.\frac{\partial^2 u_i}{\partial x_j \partial x_k}\right|^{\mu} c_k^{\mu\nu} - \left.\frac{\partial u_i}{\partial x_j}\right|^{\nu} + \left.\frac{\partial u_i}{\partial x_j}\right|^{\mu}\right)^2 = \min$$

The sum here is over all particles with which particle μ shares a Voronoi boundary. The sums of the branch vectors are approximated as integrals over spheres (3-D) or circles (2-D) and the second derivative is evaluated as

$$\left.\begin{array}{l} \displaystyle\sum_{\nu=1}^{N^\mu} c_k^{\mu\nu} \simeq 0 \\ \displaystyle\sum_{\nu=1}^{N^\mu} c_k^{\mu\nu} c_\ell^{\mu\nu} \simeq N_V^\mu \aleph^{(2,d)} \overline{c^2}^\mu \delta_{k\ell} \end{array}\right\} \rightarrow \left.\frac{\partial^2 u_i}{\partial x_j \partial x_k}\right|^{\mu} \simeq \frac{1}{N_V^\mu \aleph^{(2,d)} \overline{c^2}^\mu} \sum_{\nu=1}^{N^\mu} \left.\frac{\partial u_i}{\partial x_j}\right|^{\nu} c_k^{\mu\nu}$$

In order to respect the symmetry in the subscripts j and k the result is extended to

$$\left.\frac{\partial^2 u_i}{\partial x_j \partial x_k}\right|^{\mu} \simeq \frac{1}{2N_V^{\mu} ℵ^{(2,d)} \overline{c^2}^{\mu}} \left(\sum_{\nu=1}^{N^{\mu}} \left.\frac{\partial u_i}{\partial x_k}\right|^{\nu} c_j^{\mu\nu} + \sum_{\nu=1}^{N^{\mu}} \left.\frac{\partial u_i}{\partial x_j}\right|^{\nu} c_k^{\mu\nu} \right)$$

The connection between an incremental displacement difference $\mathbf{u}^{\nu} - \mathbf{u}^{\mu}$ and a contact force increment $\mathbf{f}^{\mu\nu}$ is the interactive 2-tensor $\mathbf{K}^{\mu\nu}$, so that

$$f_i^{\mu\nu} = K_{ij}^{\mu\nu} \left(u_j^{\nu} - u_j^{\mu} \right)$$

The interactive tensor for a frictionless interaction is composed of a 'spring constant' $k^{\mu\nu}$ and two unit vectors, that are normal to the surface across which the interaction takes place: $K_{ij}^{\mu\nu} = k^{\mu\nu} n_i^{\mu\nu} n_j^{\mu\nu}$; note that $k^{\mu\nu} = 0$ when there is no contact between μ and ν. Force equilibrium for each particle requires

$$\sum_{\nu=1}^{N^{\mu}} k^{\mu\nu} n_i^{\mu\nu} n_j^{\mu\nu} \left(u_j^{\nu} - u_j^{\mu} \right) = 0$$

The sum is over all neighbours, that is all Voronoi boundaries.

For perfectly circular or spherical particles the moment equilibrium is satisfied automatically. This is the case studied here; for particles of any other shape moment equations should be accounted for.

6.3 Mean-field approximation

Regardless of the force equilibrium, an initial estimate of the stiffness tensor can be made by using the mean strain as an approximation of the displacement difference between two neighbouring particles. For an assembly of N particles the mean strain is

$$\overline{e}_{ij} = \frac{1}{2} \overline{\left(\frac{\partial u_i}{\partial x_j} + \frac{\partial u_j}{\partial x_i} \right)} = \frac{1}{2N} \sum_{\mu=1}^{N} \left(\left.\frac{\partial u_i}{\partial x_j}\right|^{\mu} + \left.\frac{\partial u_j}{\partial x_i}\right|^{\mu} \right)$$

The average of the second derivative vanishes for a statistically uniform assembly. The mean-field imposition then makes the equilibrium equations superfluous. This is a general implication of a mean-field assumption: (some of the) equilibrium equations must be sacrificed.

The mean-field estimate for the stress increment is easily obtained (see Chapter 5, Section 5.5).

$$\bar{\sigma}_{ij} = \frac{1}{2V}\sum_{\mu}\sum_{\nu=1}^{N^\mu} f_i^{\mu\nu} c_j^{\mu\nu} = \frac{1}{2V}\sum_{\mu}\sum_{\nu=1}^{N^\mu} K_{ik}^{\mu\nu}\left(u_k^\nu - u_k^\mu\right)c_j^{\mu\nu}$$

$$= \frac{\bar{e}_{k\ell}}{2V}\sum_{\mu}\sum_{\nu=1}^{N^\mu} k^{\mu\nu} n_i^{\mu\nu} n_k^{\mu\nu} c_\ell^{\mu\nu} c_j^{\mu\nu}$$

The mean-field stiffness tensor readily follows

$$X_{ijk\ell}^{mf} = \frac{1}{2V}\sum_{\mu}\sum_{\nu=1}^{N^\mu} k^{\mu\nu} n_i^{\mu\nu} n_k^{\mu\nu} c_\ell^{\mu\nu} c_j^{\mu\nu}$$

It transpires that the mean-field estimate of the components of the stiffness is generally a very bad estimate of the stiffness tensor, one that does not stand up to experimental scrutiny. However, it is a useful object to measure the effect that fluctuations have on the mechanical response of the medium. As an initial estimate a certain amount of insight can be derived from it as well.

Another thing to note is that the form of the mean-field stiffness contains a so-called *structural sum*. These objects are defined in the List of Symbols, Section B.2.1; they are called $\mathbf{A}^{\bullet}$. The use of structural sums makes the notation much more compact, avoiding the need for lengthy sums. Care must be taken though, with the order of the subscripts. With the interaction tensor $K_{ij}^{\mu\nu} = k^{\mu\nu} n_i^{\mu\nu} n_j^{\mu\nu}$, the structural sum that comes to the fore is the second order one, and

$$A_{ijk\ell}^{\mu} = \sum_{\nu=1}^{N^\mu} K_{ij}^{\mu\nu} c_k^{\mu\nu} c_\ell^{\mu\nu} \to X_{ijk\ell}^{mf} = \frac{1}{2V}\sum_{\mu} A_{ikj\ell}^{\mu} \text{ or } X_{ijk\ell}^{mf} = \frac{1}{2v}\overline{A_{ikj\ell}^{\bullet}},$$

where the volume per particle is $v = V / N$.

By way of example a two-dimensional medium is considered, which consists of discs of near equal radius and it is assumed that all the spring constants $k^{\mu\nu}$ are equal: $k^{\mu\nu} = k$. The sum over the Voronoi boundaries can be converted to a sum over contacts using the fabric tensor **p**. For the mean-field estimate the average of this tensor is required only and a coordinate system may be chosen in which **p** is diagonal. The mean-field stiffness tensor takes the explicit form

$$\mathbf{X}^{mf} \to \frac{\pi k \bar{c}^2}{16v}\begin{pmatrix} 5\bar{p}_{11}+\bar{p}_{22} & 0 & \bar{p}_{11}+\bar{p}_{22} \\ 0 & \bar{p}_{11}+\bar{p}_{22} & 0 \\ \bar{p}_{11}+\bar{p}_{22} & 0 & \bar{p}_{11}+5\bar{p}_{22} \end{pmatrix}$$

Anisotropy in the packing is observed to lead to anisotropy in the stiffness tensor (no surprise). In a test in which the minor principal stress is kept constant a measure for the dilatancy is the ratio of the volume strain to the major principal strain, which is

$$\frac{\bar{e}_{11}+\bar{e}_{22}}{\bar{e}_{11}} = 1 - \frac{\bar{p}_{11}+\bar{p}_{22}}{\bar{p}_{11}+5\bar{p}_{22}}$$

This ratio will only ever become negative when $\bar{p}_{22}$ becomes negative. So, in order to produce a mean-field theory that exhibits volume expansion, the rather unphysical requirement of a negative contact distribution must be introduced. Nonetheless, the mean-field theory shows that anisotropy certainly helps push the ratio towards a negative value, even though it is not able to actually reproduce it.

Similarly, failure of the medium as measured by the value of the outer determinant (here normalised to X^{mf}_{1111}) cannot be reached for positive $\bar{p}_{22}$. The normalised determinant is

$$4\frac{\left(\bar{p}_{22}/\bar{p}_{11}\right)^2+6\left(\bar{p}_{22}/\bar{p}_{11}\right)+1}{\left[\left(\bar{p}_{22}/\bar{p}_{11}\right)+5\right]^2},$$

which reaches zero when $\left(\bar{p}_{22}/\bar{p}_{11}\right) = 2\sqrt{2}-3 \approx -.17$. Such a value would again be unphysical, but it does show that increased anisotropy pushes the assembly further towards the régime identified before as prone to rupture layer formation.

6.4 Perturbations to the mean-field theory

By using the single particle equilibrium equations the effects of deviations from the mean-field theory can be studied. The mean-field will be the first estimate.

The second-order derivative of the displacement field needs further investigation. The analysis of heterogeneity in a continuum, as analysed in Chapter 4 will provide a context. The latter informs on the local character of the influence of fluctuations. Therefore, an analysis of the equilibrium in the vicinity of one particle will be done. This particle is labelled μ and its neighbours are labelled ν. These particles are all in mechanical equilibrium and the kinematics must be such that the force equilibrium equations are satisfied. The second derivative can now of course not be ignored, so the equilibrium equation for a particle — μ — reads

$$\sum_{\varepsilon=1}^{N^\mu} k^{\mu\varepsilon} n_i^{\mu\varepsilon} n_j^{\mu\varepsilon} \left(\left. \frac{\partial u_j}{\partial x_k} \right|^{\mu} c_k^{\mu\varepsilon} + \frac{1}{2} \left. \frac{\partial^2 u_j}{\partial x_k \partial x_\ell} \right|^{\mu} c_k^{\mu\varepsilon} c_\ell^{\mu\varepsilon} \right) = 0$$

The rudiments of the continuum equilibrium equations are recognised and it would be good if use can be made of the calculation done for that case. In the continuum theory in the approximation of small perturbations, terms that are proportional to the average strain and proportional to the second displacement derivative are present. The basic equation, which is an expansion up to first order in the fluctuations of the strain and the stiffness, is recalled from Chapter 4, Section 4.1.

$$\bar{X}_{ijk\ell} \frac{\partial e'_{k\ell}(\mathbf{x})}{\partial x_j} + \frac{\partial X'_{ijk\ell}(\mathbf{x})}{\partial x_j} \bar{e}_{k\ell} = 0$$

The one-particle equilibrium equation and the corresponding continuum equation are different, in that the former is valid in *points* (the centre of the particles), while the latter deals with *fields*. Reconciling these two is the main problem of the analysis.

There is no problem replacing the coefficient in front of the second derivative by an average, thus

$$\frac{1}{2}\sum_{\varepsilon=1}^{N^\mu} k^{\mu\varepsilon} n_i^{\mu\varepsilon} n_j^{\mu\varepsilon} c_k^{\mu\varepsilon} c_\ell^{\mu\varepsilon} \to \overline{\frac{1}{2}\sum_{\varepsilon=1}^{N^\bullet} k^{\bullet\varepsilon} n_i^{\bullet\varepsilon} n_j^{\bullet\varepsilon} c_k^{\bullet\varepsilon} c_\ell^{\bullet\varepsilon}}$$

Now, in order to arrive at a correspondence with the continuum analysis, the term proportional to the displacement gradient needs to be expressed as a derivative. Generally, a derivative in a granular medium is obtained by a least-squares estimate. So, for a quantity q, which is defined at the centre of each particle, the derivative satisfies

$$\sum_{\nu=1}^{N^\mu}\left[\left(\frac{\partial q}{\partial x_i}\right)^\mu c_i^{\mu\nu} - q^\nu + q^\mu\right]^2 = \min$$

The sum here is over all the Voronoi boundaries of particle μ. Differentiating with respect to the components of the derivative gives

$$\sum_{\nu=1}^{N^\mu}\left[\left(\frac{\partial q}{\partial x_i}\right)^\mu c_i^{\mu\nu} - q^\nu + q^\mu\right] c_j^{\mu\nu} = 0$$

Mindful of the fact that $\sum_{\nu=1}^{N^\mu} c_j^{\mu\nu} = 0,$ the resulting expression for the derivative is

$$\left(\frac{\partial q}{\partial x_p}\right)^\mu = \left(\sum_{\nu=1}^{N^\mu} \mathbf{c}^{\mu\nu}\mathbf{c}^{\mu\nu}\right)^{-1}_{pj} \sum_{\nu=1}^{N^\mu} q^\nu c_j^{\mu\nu}$$

It is not unreasonable to approximate $\sum_{\nu=1}^{N^\mu} c_j^{\mu\nu} c_p^{\mu\nu}$ by an average: $N_V \aleph^{(2,d)} \overline{c^2} \delta_{k\ell}$, in which case the derivative becomes

$$\left(\frac{\partial q}{\partial x_p}\right)^\mu \simeq \frac{1}{N_V \aleph^{(2,d)} \overline{c^2}} \sum_{\nu=1}^{N^\mu} q^\nu c_p^{\mu\nu}$$

Now, applying this, a derivative for the structural sum is approximated as

$$\left(\frac{\partial \sum_{\kappa=1}^{N^{\bullet}} k^{\bullet\kappa} n_i^{\bullet\kappa} n_j^{\bullet\kappa} c_k^{\bullet\kappa} c_\ell^{\bullet\kappa}}{\partial x_p}\right)^{\mu} \simeq \frac{1}{N_V \aleph^{(2,d)} \overline{c^2}} \sum_{\nu=1}^{N^\mu} c_p^{\mu\nu} \sum_{\kappa=1}^{N^\nu} k^{\nu\kappa} n_i^{\nu\kappa} n_j^{\nu\kappa} c_k^{\nu\kappa} c_\ell^{\nu\kappa}$$

The sums over κ are elaborated as follows. One of the contributors is the $(\mu\nu)$ boundary. That one is taken out, the right-hand-side then takes the form

$$\frac{1}{N_V \aleph^{(2,d)} \overline{c^2}} \sum_{\nu=1}^{N^\mu} \left(k^{\nu\mu} n_i^{\nu\mu} n_j^{\nu\mu} c_k^{\nu\mu} c_\ell^{\nu\mu} c_p^{\mu\nu} + c_p^{\mu\nu} \sum_{\kappa\neq\mu}^{N^\nu} k^{\nu\kappa} n_i^{\nu\kappa} n_j^{\nu\kappa} c_k^{\nu\kappa} c_\ell^{\nu\kappa} \right)$$

The boundaries in the second sum are all *not* bordering particle μ. For these the interactive tensor is replaced by its average $\bar{k}$. Then another average contact is added and subtracted again, to give for the sum

$$\frac{\sum_{\nu=1}^{N^\mu} \left(k^{\nu\mu} n_i^{\nu\mu} n_j^{\nu\mu} c_k^{\nu\mu} c_\ell^{\nu\mu} c_p^{\mu\nu} + c_p^{\mu\nu} \sum_{\kappa=1}^{N^\nu} \bar{k} n_i^{\nu\kappa} n_j^{\nu\kappa} c_k^{\nu\kappa} c_\ell^{\nu\kappa} - \bar{k} n_i^{\nu\mu} n_j^{\nu\mu} c_k^{\nu\mu} c_\ell^{\nu\mu} c_p^{\mu\nu} \right)}{N_V \aleph^{(2,d)} \overline{c^2}}$$

Using $\sum_{\nu=1}^{N^\mu} c_j^{\mu\nu} = 0$, approximately the sum over any odd string of coordinate vector components may be neglected. This leaves

$$\left(\frac{\partial \sum_{\kappa=1}^{N^{\bullet}} k^{\bullet\kappa} n_i^{\bullet\kappa} n_j^{\bullet\kappa} c_k^{\bullet\kappa} c_\ell^{\bullet\kappa}}{\partial x_p}\right)^{\mu} \simeq \frac{1}{N_V \aleph^{(2,d)} \overline{c^2}} \sum_{\nu=1}^{N^\mu} \left(k^{\nu\mu} - \bar{k}\right) n_i^{\nu\mu} n_j^{\nu\mu} c_k^{\nu\mu} c_\ell^{\nu\mu} c_p^{\mu\nu}$$

Multiplying with $\delta_{\ell p}$ and summing over p then yields

$$\left(\frac{\partial \sum_{\kappa=1}^{N^{\bullet}} k^{\bullet\kappa} n_i^{\bullet\kappa} n_j^{\bullet\kappa} c_k^{\bullet\kappa} c_\ell^{\bullet\kappa}}{\partial x_\ell}\right)^{\mu} \simeq \frac{1}{N_V \aleph^{(2,d)}} \sum_{\nu=1}^{N^\mu} \left(k^{\nu\mu} - \bar{k}\right) n_i^{\nu\mu} n_j^{\nu\mu} c_k^{\nu\mu} \tag{6.1}$$

The centre of the particle coordinate system is chosen in such a way that $\sum_{\nu=1}^{N^{\mu}} c_j^{\mu\nu} = 0$. While that does not immediately imply that the sum over the three string is also zero, the majority of the fluctuation is going to be due to the variability in contact properties and in the right-hand side of Equation (5.1) the odd structural sum is recognised. It is then suggested that the one-particle equilibrium equation is rewritten as

$$\sum_{\varepsilon=1}^{N^{\mu}} k^{\mu\varepsilon} n_i^{\mu\varepsilon} n_j^{\mu\varepsilon} \left(\left.\frac{\partial u_j}{\partial x_k}\right|^{\mu} c_k^{\mu\varepsilon} + \frac{1}{2} \left.\frac{\partial^2 u_j}{\partial x_k \partial x_\ell}\right|^{\mu} c_k^{\mu\varepsilon} c_\ell^{\mu\varepsilon} \right) = 0 \rightarrow$$

$$N_V \aleph^{(2,d)} \overline{\frac{\partial u_j}{\partial x_k}} \left(\frac{\partial \sum_{\kappa=1}^{N^{\bullet}} k^{\bullet\kappa} n_i^{\bullet\kappa} n_j^{\bullet\kappa} c_k^{\bullet\kappa} c_\ell^{\bullet\kappa}}{\partial x_\ell} \right)^{\mu} \tag{6.2}$$

$$+ \frac{1}{2} \left.\frac{\partial^2 u_j}{\partial x_k \partial x_\ell}\right|^{\mu} \overline{\sum_{\varepsilon=1}^{N^{\bullet}} k^{\bullet\varepsilon} n_i^{\bullet\varepsilon} n_j^{\bullet\varepsilon} c_k^{\bullet\varepsilon} c_\ell^{\bullet\varepsilon}} \simeq 0$$

This equation has the same structure as the continuum equilibrium equation and the solution is readily taken over. For convenience of notation the fluctuating field is denoted with a prime

$$A'_{ijk\ell} = \left(\sum_{\kappa=1}^{N^{\bullet}} k^{\bullet\kappa} n_i^{\bullet\kappa} n_j^{\bullet\kappa} c_k^{\bullet\kappa} c_\ell^{\bullet\kappa} \right)'$$

This notation will enable a compact representation of the solution of Equation (6.2).

6.5 Solution in two dimensions

In the following the structure of the problem is illustrated in a 2-D isotropic example. For this case $\aleph^{(2,2)} = \frac{1}{2}$. The solution for the displacement (as obtained in the continuum theory) is in Fourier transformed variables

$$\hat{u}_a = iN_V P_{ai}^{-1} k_j \hat{A}'_{ijk\ell} \overline{e}_{k\ell},$$

$$\text{where } P_{ai}^{-1} = -\frac{\left(\overline{\lambda}+\overline{\mu}\right)}{\overline{\mu}\left(\overline{\lambda}+2\overline{\mu}\right)k^4} k_a k_i + \frac{1}{\overline{\mu} k^2}\delta_{ai}$$

For the present problem $\overline{\lambda} = \overline{\mu}$, as can be seen from the mean-field moduli. (However, the slightly more general problem of $\overline{\lambda} \neq \overline{\mu}$ can easily be treated at the same time.)

In the spatial domain the fluctuation in the displacement field is

$$u'_a(\mathbf{x}) = \frac{i}{(2\pi)^2}\int d_2 k e^{i\mathbf{k}.\mathbf{x}} P_{ai}^{-1} k_j \int d_2 y A'_{ijk\ell}(\mathbf{y}) e^{-i\mathbf{k}.\mathbf{y}} \overline{e}_{k\ell} \tag{6.3}$$

The fluctuations $\mathbf{A}'$ appear here as a *continuous field*, but they are defined in a granular medium as quantities on *points* (the centres of the particles). Reconciling these two notions is done by putting forward a continuous field and — noting that only the value in the vicinity of particle μ is required — letting this field decay away from the centre of the particle. Furthermore, the assumption is made that the fluctuations are purely radial. The integral over the angle is easily done

$$\int d_2 y A'_{ijk\ell}(\mathbf{y}) e^{i\mathbf{k}.\mathbf{y}} = 2\pi \int_0^\infty dy y J_0(ky) A'_{ijk\ell}(y)$$

Let the radial dependence be of the form $A'_{ijk\ell}(0)\exp\left(-y^2/\widehat{a}^2\right)$, where $\widehat{a}$ is an as yet adjustable parameter with the dimension of a length, then (see Appendix, Section A.6.1)

$$\begin{aligned}\int d_2 y A'_{ijk\ell}(\mathbf{y}) e^{i\mathbf{k}.\mathbf{y}} &= 2\pi \int_0^\infty dy y J_0(ky) A'_{ijk\ell}(y) \\ &= \pi \widehat{a}^2 \exp\left(-\frac{1}{4}\widehat{a}^2 k^2\right) A'_{ijk\ell}(0)\end{aligned}$$

Consequently,

$$u'_a(\mathbf{x}) = \frac{i\hat{a}^2 N_V}{4\pi} \int_0^{2\pi} d\psi e^{i\mathbf{k}.\mathbf{x}} \int_0^{\infty} dkkP_{ai}^{-1} k_j \exp\left(-\frac{1}{4}\hat{a}^2 k^2\right) A'_{ijk\ell}(0)\overline{e}_{k\ell}$$

These two integrals are worked out in Appendix, Section A.6.1, first the integral over ψ

$$u'_a(\mathbf{x}) = -\frac{1}{2}\hat{a}^2 N_V p_0 m_j \delta_{ai} \int_0^{\infty} dk \exp\left(-\frac{1}{4}\hat{a}^2 k^2\right) J_1(kx) A'_{ijk\ell}(0)\overline{e}_{k\ell}$$
$$-\hat{a}^2 N_V p_1 \int_0^{\infty} dk \exp\left(-\frac{1}{4}\hat{a}^2 k^2\right)\left(\frac{1}{8}\left[J_1(kx) + J_3(kx)\right]\left(m_i\delta_{aj} + m_j\delta_{ai} + m_a\delta_{ij}\right)\right.$$
$$\left. -\frac{1}{2} J_3(kx) m_i m_j m_a\right) A'_{ijk\ell}(0)\overline{e}_{k\ell},$$

with the following

$$p_0 = \frac{1}{\overline{\mu}};\ p_1 = -\frac{\left(\overline{\lambda} + \overline{\mu}\right)}{\overline{\mu}\left(\overline{\lambda} + 2\overline{\mu}\right)}$$

and defining the two functions

$$\begin{aligned} S_1\left(\frac{x}{\hat{a}}\right) &\equiv \hat{a}\int_0^{\infty} dk \exp\left(-\frac{1}{4}\hat{a}^2 k^2\right) J_1(kx) = \sqrt{\pi}\exp\left(-\frac{x^2}{2\hat{a}^2}\right) I_{\frac{1}{2}}\left(\frac{x^2}{2\hat{a}^2}\right) \\ S_3\left(\frac{x}{\hat{a}}\right) &\equiv \hat{a}\int_0^{\infty} dk \exp\left(-\frac{1}{4}\hat{a}^2 k^2\right) J_3(kx) = \sqrt{\pi}\exp\left(-\frac{x^2}{2\hat{a}^2}\right) I_{\frac{3}{2}}\left(\frac{x^2}{2\hat{a}^2}\right) \end{aligned}, \qquad (6.4)$$

the integrals over k are evaluated, to give the result

$$u'_a(\mathbf{x}) = -\frac{\hat{a}N_V}{2}\left[p_1\left(\left[S_1\left(\frac{x}{\hat{a}}\right) + S_3\left(\frac{x}{\hat{a}}\right)\right]\frac{m_i\delta_{aj} + m_j\delta_{ai} + m_a\delta_{ij}}{4}\right)\right] A'_{ijk\ell}(0)\overline{e}_{k\ell}$$
$$-\frac{\hat{a}N_V}{2} p_1 S_3\left(\frac{x}{\hat{a}}\right) m_i m_j m_a A'_{ijk\ell}(0)\overline{e}_{k\ell} - \frac{\hat{a}N_V}{2} p_0 m_j \delta_{ai} S_1\left(\frac{x}{\hat{a}}\right) A'_{ijk\ell}(0)\overline{e}_{k\ell}$$

This is the fluctuating part of the displacement due to a fluctuation in the structural sum. The displacement so calculated gives rise to an extra term in the stress, just like the continuum analysis. The extra stress due to the fluctuating field is expressed as a fraction of the mean-field shear modulus $\bar{\mu}$ (in this way the proportionality factor $(2V)^{-1}$ does not have to be included in the calculation). The required S-functions are plotted in Fig. 6.1.

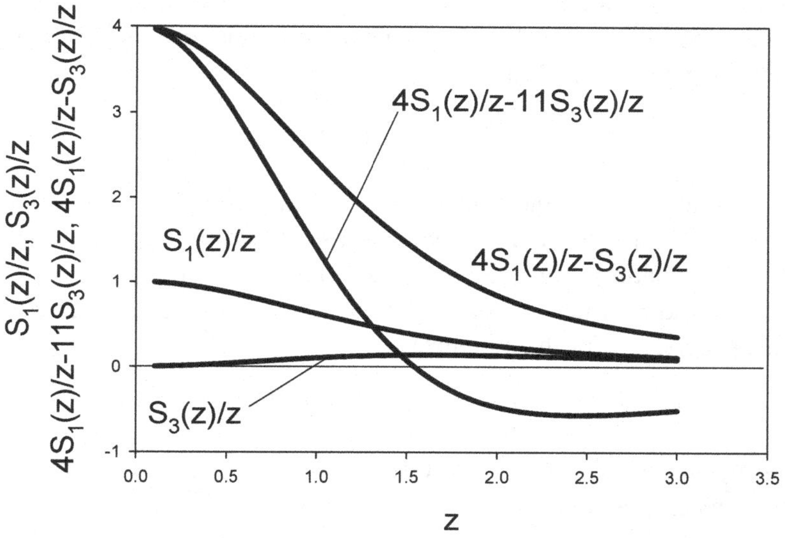

Figure 6.1. The functions $S_1(z)/z$ and $S_3(z)/z$.

The mean modulus is

$$\bar{\mu} = \bar{c}^2 \bar{p}\bar{k} \int_0^{2\pi} n_1^2 n_2^2 d\varphi = \frac{\pi}{4} \bar{c}^2 \bar{p}\bar{k}$$

The contribution to the relative stress due to fluctuations is

$$\frac{\overline{\sigma'_{rs}}}{\bar{\mu}} = \frac{1}{\bar{\mu}} \overline{\sum_{\nu=1}^{N^\mu} \left(k^{\mu\nu} - \bar{k}\right) m_r^{\mu\nu} m_a^{\mu\nu} u_a'^{\mu\nu} c_s^{\mu\nu}}$$

The displacement fluctuation $\mathbf{u}'^{\mu\nu}$ is identified with the displacement fluctuation as calculated above with $\mathbf{u}'\left(\mathbf{c}^{\mu\nu}\right)$. For disc-shaped particles the term $\mathbf{A}'(0)$ is equal to

$$\bar{c}^2\left(\sum_{\kappa=1}^{N^\bullet} k^{\bullet\kappa}\mathbf{n}^{\bullet\kappa}\mathbf{n}^{\bullet\kappa}\mathbf{n}^{\bullet\kappa}\mathbf{n}^{\bullet\kappa}\right)' \simeq \bar{c}^2\left(\sum_{\kappa=1}^{N^\bullet}\left(k^{\bullet\kappa}-\bar{k}\right)\mathbf{n}^{\bullet\kappa}\mathbf{n}^{\bullet\kappa}\mathbf{n}^{\bullet\kappa}\mathbf{n}^{\bullet\kappa}\right).$$

Therefore, the result depends on the correlation in the fluctuations in the contact point distribution. In order to get an idea how the latter becomes manifest, assume a quadratic contact distribution on each individual particle. This takes the form

$$p^{\mu}(\varphi) = p_{ij}^{\mu} n_i(\varphi) n_j(\varphi)$$

The fluctuations in the contact distribution are obtained by removing the average $\bar{p}$. In addition the fluctuations are partitioned in an isotropic part P' (indicating a fluctuation in the number of contacts per particle) and a remainder, which serves as a measure of anisotropy. Thus (leaving out the superscript for the moment)

$$p_{11} = \bar{p} + P' + p'_{11};\ p_{22} = \bar{p} + P' + p'_{22};\ p_{12} = p_{21} = p'_{12}$$

Working out the correlations results in the relative contributions of the Lamé constants as follows:

$$\begin{aligned}\frac{\lambda'}{\bar{\lambda}}\frac{\bar{c}}{\hat{a}}\frac{1}{N_V} = {} & \frac{11S_3 - 4S_1}{6\bar{p}^2}\left[\overline{(P')^2} + \overline{P'p'_{11}} + \overline{P'p'_{22}}\right] \\ & - \frac{S_1}{12\bar{p}^2}\left[3\overline{(p'_{11})^2} + 3\overline{(p'_{22})^2} + 8\overline{(p'_{12})^2} + 2\overline{p'_{11}p'_{22}}\right] \\ & + \frac{S_3}{24\bar{p}^2}\left[9\overline{(p'_{11})^2} + 9\overline{(p'_{22})^2} + 4\overline{(p'_{12})^2} + 26\overline{p'_{11}p'_{22}}\right]\end{aligned}$$

$$\begin{aligned}\frac{\mu'}{\bar{\mu}}\frac{\bar{c}}{\hat{a}}\frac{1}{N_V} = {} & \frac{S_3 - 4S_1}{6\bar{p}^2}\left[\overline{(P')^2} + \overline{P'p'_{11}} + \overline{P'p'_{22}}\right] \\ & - \frac{S_1}{6\bar{p}^2}\left[\overline{(p'_{11})^2} + \overline{(p'_{22})^2} + 2\overline{(p'_{12})^2} + 2\overline{p'_{11}p'_{22}}\right] \\ & + \frac{S_3}{24\bar{p}^2}\left[\overline{(p'_{11})^2} + \overline{(p'_{22})^2} + 12\overline{(p'_{12})^2} + 2\overline{p'_{11}p'_{22}}\right]\end{aligned}$$

The functions S_1 and S_3 are evaluated in the point $\bar{c}/\hat{a}$; these, divided by $\bar{c}/\hat{a}$, are plotted in Fig. 6.1.

The question then is what happens to the S-functions? In the graph — Fig. 6.1. — it is observed that these are quite sensitive functions of the argument $z=\bar{c}/\hat{a}$, in other words, on the choice of the adjustable length scale parameter $\hat{a}$.

6.6 Connectivity in a granular medium

Insight in the background to this parameter $\hat{a}$ can be obtained by a little analysis, involving the heterogeneity of and the connectedness in the granular assembly.

The analysis is carried out as follows. Consider a contact interactive parameter, such as the contact stiffness, which is generically called q. It obviously has the property $q^{\mu\nu}=q^{\nu\mu}$ for particle pairs that share a Voronoi boundary. Now evaluate the cross correlation between the contact parameter and the fluctuation of the structural sum of the neighbouring particles, in other words investigate the expression

$$\sum_{\mu}\sum_{\nu=1}^{N^{\mu}}\left[q^{\mu\nu}\left(\sum_{\varepsilon=1}^{N^{\nu}}q^{\nu\varepsilon}-N_{\nu}\bar{q}\right)\right]$$

The cross correlation is appropriate for the investigation of the influence function $\exp\left(-x^2/\hat{a}^2\right)$. If the mean distance between particle μ and ν is represented by $\bar{x}$, then *in the correlation* the fluctuation of the structural sum is *on average* represented by

$$\sum_{\varepsilon=1}^{N^{\nu}}q^{\nu\varepsilon}-N_{\nu}\bar{q}\rightarrow\left(\sum_{\pi=1}^{N^{\mu}}q^{\mu\pi}-N_{\nu}\bar{q}\right)\exp\left(-\bar{x}^2/\hat{a}^2\right)$$

So that the cross correlation takes the form

$$\sum_{\mu}\sum_{\nu=1}^{N^{\mu}}\left[q^{\mu\nu}\left(\sum_{\pi=1}^{N^{\mu}}q^{\mu\pi}-N_{\nu}\bar{q}\right)\exp\left(-\bar{x}^2/\hat{a}^2\right)\right]$$

Call the local average of the interactive parameter $\overline{q}^{\mu}$, then this expression is rewritten as

$$N_V^2 \sum_{\mu} \left[\left(\overline{q}^{\mu}\right)^2 - \overline{q}\overline{q}^{\mu} \right] \exp\left(-\overline{x}^2 / \hat{a}^2\right)$$

On the other hand the cross correlation is obtained by direct calculation

$$\sum_{\mu} \sum_{\nu=1}^{N^{\mu}} \left[q^{\mu\nu} \left(\sum_{\varepsilon=1}^{N^{\nu}} q^{\nu\varepsilon} - N_V \overline{q} \right) \right] = \sum_{\mu} \sum_{\nu=1}^{N^{\mu}} \left[q^{\mu\nu} \left(q^{\mu\nu} + \sum_{\varepsilon \neq \mu}^{N^{\nu}} q^{\nu\varepsilon} - N_V \overline{q} \right) \right]$$

Referring back to the analysis done on the correlation in contact properties in Chapter 5, Section 5.3 (especially note the spectral intensity function depicted in Chapter 5, Fig. 5.4), it is not unreasonable to neglect the cross correlate compared to the auto correlate; this approximation leads to

$$\begin{aligned}
\sum_{\mu} \sum_{\nu=1}^{N^{\mu}} \left[q^{\mu\nu} \left(q^{\mu\nu} + \sum_{\varepsilon \neq \mu}^{N^{\nu}} q^{\nu\varepsilon} \right) \right] &= \sum_{\mu} \sum_{\nu=1}^{N^{\mu}} \left[q^{\mu\nu} \left(q^{\mu\nu} - \overline{q} + \overline{q} + \sum_{\varepsilon \neq \mu}^{N^{\nu}} q^{\nu\varepsilon} - N_V \overline{q} \right) \right] \\
&\approx \sum_{\mu} \sum_{\nu=1}^{N^{\mu}} \left[q^{\mu\nu} \left(q^{\mu\nu} - \overline{q} \right) \right] = \sum_{\mu} \left(\sum_{\nu=1}^{N^{\mu}} \left(q^{\mu\nu}\right)^2 - N_V \overline{q}^{\mu} \overline{q} \right) \\
&= N_V \sum_{\mu} \left(\overline{\left(q^2\right)}^{\mu} - \overline{q}\overline{q}^{\mu} \right)
\end{aligned}$$

Now, the interactive parameter may be written as a local average plus fluctuations

$$q^{\mu\nu} = \overline{q}^{\mu} + q'^{\mu\nu}$$

So that

$$\begin{aligned}
N_V \sum_{\mu} \left(\overline{\left(q^2\right)}^{\mu} - \overline{q}\overline{q}^{\mu} \right) &= \sum_{\mu} \left[\sum_{\nu=1}^{N^{\mu}} \left(\overline{q}^{\mu} + q'^{\mu\nu} \right)^2 - N_V \overline{q}\overline{q}^{\mu} \right] \\
&= \sum_{\mu} \left(N_V \left(\overline{q}^{\mu}\right)^2 + \sum_{\nu=1}^{N^{\mu}} \left(q'^{\mu\nu}\right)^2 - N_V \overline{q}\overline{q}^{\mu} \right)
\end{aligned}$$

Combining both evaluations yields

$$\exp\left(-\overline{x}^2 / \widehat{a}^2\right) \simeq \frac{\sum_{\mu}\left(N_V\left(\overline{q}^{\mu}\right)^2 + \sum_{\nu=1}^{N^{\mu}}\left(q'^{\mu\nu}\right)^2 - N_V\overline{q}\overline{q}^{\mu}\right)}{N_V^2\sum_{\mu}\left[\left(\overline{q}^{\mu}\right)^2 - \overline{q}\overline{q}^{\mu}\right]}$$

$$= \frac{\left(\overline{\left(\overline{q}^{\bullet}\right)^2} + \overline{\overline{\left(q'\right)^2}^{\bullet}} - \overline{q}^2\right)}{N_V\left[\overline{\left(\overline{q}^{\bullet}\right)^2} - \overline{q}^2\right]}$$

The case of non-zero fluctuations is studied in an example. Let the contact parameter q be the contact itself.

Two cases are considered:

1. A contact parameter that equals unity when there is a Voronoi boundary with a contact and zero when it pertains to a Voronoi boundary that does not correspond to a contact.
2. A contact parameter that pertains to the case where the average contact value vanishes. This case is of particular interest, as the use of the exponential influence function was first introduced for fluctuating structural sums.

The first case is investigated in the following manner. For a particle μ with N_c^{μ} contacts the value of the contact parameter $q^{\mu\nu}$ equals 1 or 0. If there are N_V Voronoi boundaries the one-particle average is

$$\overline{q}^{\mu} = \frac{N_c^{\mu}}{N_V}$$

And the fluctuation at the contact

$$q'^{\mu\nu} = \begin{cases} 1 - \dfrac{N_c^{\mu}}{N_V}(contact) \\ -\dfrac{N_c^{\mu}}{N_V}(no\,contact) \end{cases}$$

Thus the quadratic average of the fluctuations is

$$\frac{1}{N_V}\sum_{\nu=1}^{N^{\mu}}\left(q'^{\mu\nu}\right)^2=\frac{1}{N_V}\left[N_c^{\mu}\left(1-\frac{N_c^{\mu}}{N_V}\right)^2+\left(N_V-N_c^{\mu}\right)\left(\frac{N_c^{\mu}}{N_V}\right)^2\right]$$

Writing the number of contacts as a deviation of the assembly-average N_c: $N_c^{\mu}=N_c+\delta N_c^{\mu}$ and define $f_c=\left(\delta N_c\right)^2/N_c^2$, it follows that

$$\exp\left(-\overline{x}^2/\hat{a}^2\right)\approx\frac{\left(\overline{\left(\overline{q}^{\bullet}\right)^2}+\overline{\overline{\left(q'\right)^2}^{\bullet}}-\overline{q}^2\right)}{N_V\left[\overline{\left(\overline{q}^{\bullet}\right)^2}-\overline{q}^2\right]}=\frac{N_V-N_c}{N_VN_cf_c}$$

The second case is at first sight very different; it assigns to the contact the values

$$q^{\mu\nu}=\begin{cases}1-\dfrac{N_c}{N_V}(contact)\\-\dfrac{N_c}{N_V}(no\,contact)\end{cases}$$

Consequently, the one-particle average is

$$\overline{q}^{\mu}=\frac{N_c^{\mu}}{N_V}\left(1-\frac{N_c}{N_V}\right)-\frac{N_V-N_c^{\mu}}{N_V}\frac{N_c}{N_V}=\frac{N_c^{\mu}-N_c}{N_V}$$

This, of course vanishes when $N_c^{\mu}=N_c$. The fluctuations are

$$q'^{\mu\nu}=\begin{cases}1-\dfrac{N_c}{N_V}-\overline{q}^{\mu}(contact)\\-\dfrac{N_c}{N_V}-\overline{q}^{\mu}(no\,contact)\end{cases}$$

The answer is the same as before

$$\exp\left(-\overline{x}^2 / \hat{a}^2\right) \simeq \frac{N_V - N_c}{N_V N_c f_c}$$

The result of this formula is plotted in Fig. 6.2. for the choice $N_V = 6$ and $\overline{x}/\overline{c} = 1.18$ for three values of the contact number. The first conclusion is that if the value of f_c is larger, the value of $\overline{c}/\hat{a}$ is in the neighbourhood of unity. For smaller fluctuations the value goes up and can pretty well double. The consequence of this is that when there are large fluctuations in the number of contacts, the material is also more sensitive to fluctuations, as the S-functions decline with increasing $\overline{c}/\hat{a}$. Also, for larger total numbers of particles the sensitivity to fluctuations goes up, however when there are larger numbers of particles, the value of f_c tends to be smaller in practice. Overall, the outcome is very reasonable. Note that while $N_c \neq N_V$ there *must* be fluctuations.

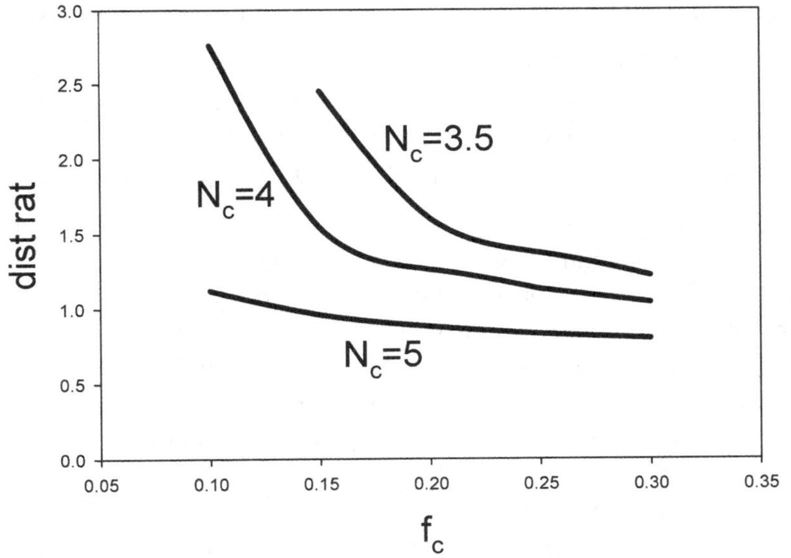

Figure 6.2. The distance ratio $\overline{c}/\hat{a}$ for $N_V = 6$ and $N_c = 3.5, 4.0, 5.0$.

The conclusion is that the 'smearing-out' as represented by the exponential influence function is associated with the connectivity in the medium — manifest via the property $q^{\mu\nu} = q^{\nu\mu}$ — and the fabric heterogeneity. The actual functional form — the e-power — cannot be ascertained in this way and the choice of an exponential has been made purely for mathematical convenience.

6.7 Estimates of the correction to the moduli due to heterogeneity

Looking first at the shear modulus and considering the isotropic part of the fluctuations only, it is observed that with $N_V \simeq 6$, this quantity is reduced by a factor

$$\left[4S_1\left(\frac{\overline{c}}{\widehat{a}}\right)-S_3\left(\frac{\overline{c}}{\widehat{a}}\right)\right]\frac{\widehat{a}}{\overline{c}}\frac{\overline{(P')^2}}{\overline{p}^2}$$

For a fairly dense packing, for example the one demonstrated in Fig. 5.1, the ratio $\overline{(P')^2}/\overline{p}^2$ has been determined to be approximately 0.35 ($\overline{N}_c \simeq 4.1$). For looser packings the ratio tends to become greater. A plausible value is in the range $0.5 < \overline{c}/\widehat{a} < 1.0$. In this range the correction to the shear modulus due to fluctuations of the average number of contacts lies in the range $0.9 < \left|\overline{\mu} - \overline{\mu}^{(mf)}\right| / \overline{\mu}^{(mf)} < 1.2$. The additional correction due to anisotropic effects may be greater. What this shows is that the effective shear modulus, that is the mean field value and the reduction due to packing fluctuations together, collapses to zero.

Referring to the analysis of rupture layer formation, a shear modulus that approaches a zero value leads to major instability in the assembly as a whole. So, packings with a mean number of contacts less than, say, $N_c \simeq 4$ cannot stably exist. It must be pointed out that assemblies with contacting particles with normal interactions only are in reality highly unlikely. Particles in contact will have a tangential contact stiffness as well (this case is treated in Chapter 7). It is, however, possible to create such assemblies in a computer simulation and a very low shear modulus has indeed been reported: [Magnanimo *et al.*, 2008]. For the cakes in which small sub-micron particles are packed that experience the normal interaction only, high contact numbers and low fluctuations are necessary for a stable conformation.

It is tempting to compare this estimate with the analysis done on isostatics. It was found that for perfectly round particles with frictionless interaction, $N_c = 4$ is the minimum number of permissible contacts. The heterogeneity analysis does not account for numbers of equations and unknowns and therefore the only way in which it can predict stability of the assembly is via the rupture layer analysis applied with the homogenised moduli. Yet, there appears to be a correspondence.

Turning now to the other Lamé constant. Using again the isotropic part of the fluctuations as a first order estimate the reduction amounts to

$$\left[4S_1\left(\frac{\overline{c}}{\widehat{a}}\right)-11S_3\left(\frac{\overline{c}}{\widehat{a}}\right)\right]\frac{\widehat{a}}{\overline{c}}\frac{\overline{(P')^2}}{\overline{p}^2}$$

In the expected range of $\overline{c}/\widehat{a}$ and using the graph in Fig. 6.1, the reduction falls in the range $0.4 < \left|\overline{\lambda} - \overline{\lambda}^{mf}\right| / \overline{\lambda}^{mf} < 1.0$, with the lower number for the highest value of $\overline{c}/\widehat{a}$. Thus this Lamé constant is much less sensitive to heterogeneity. If $\overline{c}/\widehat{a}$ should attain lower values, it might even increase somewhat.

With the effect of the fluctuations so severe, the question is whether it is useful to study higher orders in the fluctuation terms. For the analogous continuum case [Kröner, 1967] has shown a method to evaluate these for a perfectly random medium. However, the first order estimate still gives a reasonable impression of the order of magnitude of the sensitivity of the medium to fluctuations. This analysis is not followed up here. For the type of media considered here, i.e. those with a purely normal particle interaction, it is concluded that these are very sensitive to the fluctuational content of the packing properties.

The main practical application of the theory in this chapter is, as mentioned, small particles in a fluid environment. While the interaction is complex, as outlined in Chapter 1, Section 1.7, for the cases in which the interaction is repulsive there is a mechanism that minimises the heterogeneity. This mechanism relies on the elements described in this chapter, which shows that the displacement of the particles is such that it, broadly speaking, opposes the direction of the heterogeneity. In a non-linear interaction that becomes steeper the closer the particles are pressed together the magnitude of the heterogeneity will consequently become smaller due to the extra displacement $\mathbf{u}^{\bullet}$ that caused it. Therefore, in an assembly that is composed of round particles in which the packing is such that the particles are always captured in a non-linear interaction, the heterogeneity will be minimised. That implies that a mean-field analysis for such cases is not a bad approximation and has been applied successfully to describe dense filtration cakes: [Koenders and Wakeman, 1997].

References

Civan, F. (2007) *Reservoir Formation Damage: Fundamentals, Modeling, Assessment, and Mitigation*, 2nd Ed., Burlington, MA: Gulf Professional Pub., Elsevier.

Jenkins, J.T. and Koenders, M.A. (2004) The incremental response of random aggregates of round particles. *European Physics Journal E* **13** 113–123.

Koenders, M.A. (1984) A Two Dimensional Non-Homogeneous Deformation Model for Sand. PhD Thesis, University College London.

Koenders, M.A. (1987) The incremental stiffness of an assembly of particles. *Acta Mechanica* **70** 31–49.

Koenders, M.A. and Wakeman, R.J. (1997) Initial deposition of interacting particles by filtration of dilute suspensions. *AIChE Journal* **43**(4) 946–958.

Kröner, E. (1967) Elastic moduli of a perfectly disordered material. *J. Mech Phys. Solids.* **15** 319–329.

Magnanimo, V., La Ragione, L., Jenkins, J.T., Wang, P. and Makse, H.A. (2008) Characterizing the shear and bulk moduli of an idealized granular material. *Eur. Physics Letters* **81** 34006–34012.

Wakeman, R.J. and Tarleton, E.S. (1999) *Filtration: Equipment Selection, Modelling and Process Simulation.* Oxford: Elsevier Advanced Technology.

Chapter 7

Stress-Strain Relations of Granular Assemblies: Normal and Tangential Interactions

7.1 Particle spin

Spherical particles with a purely normal interaction are the exception. Especially for assemblies of contacting particles the analysis of assemblies of grains require the consideration of the particle spins, as well as the displacements of the particles. The direction of the contact is called $\mathbf{n}^{\mu\nu}$ as before. The location of the contact is measured from the centre of gravity of the particles: $\mathbf{x}^{\mu\nu}$ from the centre of particle μ and $\mathbf{x}^{\nu\mu}$ from the centre of particle ν. When an increment of deformation takes place the contact point moves by an amount $\mathbf{d}^{\mu\nu}$. This motion is partly due to the translation and partly to the rotation of the particles. The latter is denoted by the spin vector $\boldsymbol{\omega}^{\bullet}$. The contact point movement is

$$d_i^{\mu\nu} = u_i^{\nu} - u_i^{\mu} - \varepsilon_{ijk}\left(x_j^{\mu\nu}\omega_k^{\mu} - x_j^{\nu\mu}\omega_k^{\nu}\right)$$

Displacement and spin can be expanded in the neighbourhood of particle μ in a Taylor series

$$u_i\left(\mathbf{x}\right) = u_i^{\mu} + \left.\frac{\partial u_i}{\partial x_j}\right|^{\mu} x_j + \frac{1}{2}\left.\frac{\partial^2 u_i}{\partial x_j \partial x_k}\right|^{\mu} x_j x_k\,;$$

$$\omega_i\left(\mathbf{x}\right) = \omega_i^{\mu} + \left.\frac{\partial \omega_i}{\partial x_j}\right|^{\mu} x_j + \frac{1}{2}\left.\frac{\partial^2 \omega_i}{\partial x_j \partial x_k}\right|^{\mu} x_j x_k$$

The contact displacement is then approximated as

$$d_i^{\mu\nu} = \left.\frac{\partial u_i}{\partial x_j}\right|^{\mu} c_j^{\mu\nu} + \frac{1}{2}\left.\frac{\partial^2 u_i}{\partial x_j \partial x_k}\right|^{\mu} c_j^{\mu\nu} c_k^{\mu\nu} - \varepsilon_{ijk}\left(x_j^{\mu\nu} - x_j^{\nu\mu}\right)\omega_k^{\mu}$$
$$+ \varepsilon_{ijk}\left.\frac{\partial \omega_k}{\partial x_\ell}\right|^{\mu} x_j^{\nu\mu} c_\ell^{\mu\nu} + \frac{1}{2}\varepsilon_{ijk}\left.\frac{\partial^2 \omega_k}{\partial x_\ell \partial x_m}\right|^{\mu} x_j^{\nu\mu} c_\ell^{\mu\nu} c_m^{\mu\nu},$$

where the branch vector $\mathbf{c}^{\mu\nu} = \mathbf{x}^{\mu\nu} - \mathbf{x}^{\nu\mu}$.

The displacement gradient can be split in a symmetric and an anti-symmetric part. Collecting the first and the third terms together shows

$$\left.\frac{\partial u_i}{\partial x_j}\right|^{\mu} c_j^{\mu\nu} - \varepsilon_{ijk}\left(x_j^{\mu\nu} - x_j^{\nu\mu}\right)\omega_k^{\mu} = e_{ij}^{\mu} c_j^{\mu\nu} + \frac{1}{2}\left(\left.\frac{\partial u_i}{\partial x_j}\right|^{\mu} - \left.\frac{\partial u_j}{\partial x_i}\right|^{\mu}\right) c_j^{\mu\nu} - \varepsilon_{ijk} c_j^{\mu\nu}\omega_k^{\mu}$$

Any anti-symmetric tensor can be written as

$$\frac{1}{2}\left(\left.\frac{\partial u_i}{\partial x_j}\right|^{\mu} - \left.\frac{\partial u_j}{\partial x_i}\right|^{\mu}\right) = \varepsilon_{ijk}\vartheta_k^{\mu},$$

where $\boldsymbol{\vartheta}^{\mu}$ is the local displacement gradient rotation vector. The contact displacement is observed to contain only the difference of the local spin vector and the local displacement rotation vector. It makes sense, therefore, to work with the variable $\boldsymbol{\eta} \equiv \boldsymbol{\omega} - \boldsymbol{\vartheta}$ instead of $\boldsymbol{\omega}$, directly absorbing the frame rotation term. It follows than that the contact displacement is insensitive to the local displacement gradient rotation vector. This is as expected: the rigid body rotation has no influence on the contact displacement vector, which determines the interactive properties.

The second displacement derivative is now considered together with the first spin derivative. The question is what impact the frame rotation will have on this combination.

$$\frac{1}{2}\frac{\partial^2 u_i}{\partial x_j \partial x_k}\bigg|^{\mu} c_j^{\mu\nu} c_k^{\mu\nu} + \varepsilon_{ijk}\frac{\partial \omega_k}{\partial x_\ell}\bigg|^{\mu} x_j^{\nu\mu} c_\ell^{\mu\nu}$$

$$= \frac{1}{4}\left(\frac{\partial}{\partial x_j}\frac{\partial u_i}{\partial x_k} + \frac{\partial}{\partial x_k}\frac{\partial u_i}{\partial x_j}\right)\bigg|^{\mu} c_j^{\mu\nu} c_k^{\mu\nu} + \varepsilon_{ijk}\frac{\partial \omega_k}{\partial x_\ell}\bigg|^{\mu} x_j^{\nu\mu} c_\ell^{\mu\nu}$$

$$= \frac{1}{4}\left(\frac{\partial}{\partial x_j}\left(e_{ik} + \varepsilon_{ik\ell}\vartheta_\ell^{\bullet}\right) + \frac{\partial}{\partial x_k}\left(e_{ij} + \varepsilon_{ij\ell}\vartheta_\ell^{\bullet}\right)\right)\bigg|^{\mu} c_j^{\mu\nu} c_k^{\mu\nu} + \varepsilon_{ijk}\frac{\partial \omega_k}{\partial x_\ell}\bigg|^{\mu} x_j^{\nu\mu} c_\ell^{\mu\nu}$$

$$= \frac{1}{2}\left(\frac{\partial e_{ik}}{\partial x_j}\right)\bigg|^{\mu} c_j^{\mu\nu} c_k^{\mu\nu} + \frac{1}{2}\varepsilon_{ik\ell}\left(\frac{\partial \vartheta_\ell^{\bullet}}{\partial x_j}\right)\bigg|^{\mu} c_j^{\mu\nu} c_k^{\mu\nu} + \varepsilon_{ij\ell}\frac{\partial \omega_\ell}{\partial x_k}\bigg|^{\mu} x_j^{\nu\mu} c_k^{\mu\nu}$$

On average $x_j^{\nu\mu} = -\tfrac{1}{2}c_j^{\mu\nu}$ for contacting particles, but on a particle scale (especially for strongly hetero-disperse samples) this is not necessarily the case. So, as far as the second derivative is concerned, it is an approximation to absorb the frame rotation in the particle spin. Otherwise, the only spin variable that needs to be taken into account is $\boldsymbol{\omega} - \boldsymbol{\vartheta}$ and the only deformation gradient measure is the strain.

It is doubtful if higher order derivatives make a useful contribution, see again the table comparing the number of particles that can be described with the number of coefficients from the Taylor series, Chapter 6, Section 6.2.

7.2 The interaction and the quasi-static equilibrium equations

A contact force increment is linked to a contact displacement increment. The relation between these two is called the (contact) interaction. The interaction takes the form

$$f_i^{\mu\nu} = K_{ij}^{\mu\nu} d_j^{\mu\nu}$$

The contact direction $\mathbf{n}^{\mu\nu}$ is perpendicular to the solid surfaces and therefore the interactive tensor must be sensitive to the anisotropy that is associated with the direction of these surfaces. For spheres or discs this is the only geometrical parameter that enters into the analysis. The tensor should be decomposed in a normal and tangential part. For typically frictional interactions the tensor $\mathbf{K}^{\bullet\bullet}$ will also depend on the contact force $\mathbf{F}^{\bullet\bullet}$.

At this point no specifics regarding the interaction are put forward yet. It is sufficient to note that there is a direction-dependent tensor $\mathbf{K}^{\bullet\bullet}$.

The equilibrium equations for each particle require the sum of forces and the sum of moments to vanish:

$$\sum_{\nu=1}^{N^\mu} f_i^{\mu\nu} = 0 \rightarrow \sum_{\nu=1}^{N^\mu} K_{ij}^{\mu\nu}\left[u_j^\nu - u_j^\mu - \varepsilon_{jk\ell}\left(x_k^{\mu\nu}\omega_\ell^\mu - x_k^{\nu\mu}\omega_\ell^\nu\right)\right] = 0$$

$$\varepsilon_{ijk}\sum_{\nu=1}^{N^\mu} f_j^{\mu\nu} x_k^{\mu\nu} = 0 \rightarrow \varepsilon_{ijk}\sum_{\nu=1}^{N^\mu} K_{j\ell}^{\mu\nu} x_k^{\mu\nu}\left[u_\ell^\nu - u_\ell^\mu - \varepsilon_{\ell mn}\left(x_m^{\mu\nu}\omega_n^\mu - x_m^{\nu\mu}\omega_n^\nu\right)\right] = 0$$

For an assembly of N particles in d dimensions there are dN force equilibrium equations and $(2d-3)N$ moment equilibrium equations, matched by equal numbers of displacement and particle spin increments. The solution to these equations requires the specification of the mean strain increment. For a statistically homogeneous (not necessarily isotropic) assembly this should be the only condition that is imposed.

7.3 Mean-field stiffness estimate

The mean-field approximation is achieved by letting neighbouring particles move according to the mean strain $\overline{\mathbf{e}}$, which prescribes the displacement increments: $u_\ell^\nu - u_\ell^\mu = \overline{e}_{\ell m} c_m^{\mu\nu}$. However, a spin increment cannot be imposed as an assembly average. This is one of the complexities that arises when the analysis is extended to non-normal contact effects. The moment equation is employed to arrive at an estimate of the particle spin $\boldsymbol{\eta}^{\bullet}$, derived from the mean-field displacement. First a measure of the local spin is obtained from

$$\varepsilon_{ijk}\sum_{\nu=1}^{N^\mu} K_{j\ell}^{\mu\nu} x_k^{\mu\nu}\left[\bar{e}_{\ell m} c_m^{\mu\nu} - \varepsilon_{\ell mn}\eta_n^\mu\left(x_m^{\mu\nu} - x_m^{\nu\mu}\right)\right] = 0 \rightarrow$$

$$\varepsilon_{ijk}\varepsilon_{\ell mn}\eta_n^\mu \sum_{\nu=1}^{N^\mu} K_{j\ell}^{\mu\nu} x_k^{\mu\nu} c_m^{\mu\nu} = \varepsilon_{ijk}\bar{e}_{\ell m}\sum_{\nu=1}^{N^\mu} K_{j\ell}^{\mu\nu} x_k^{\mu\nu} c_m^{\mu\nu}$$

Here, the spin gradient, which controls the difference in particle spins between contacting neighbours has been neglected. The idea is that the imposed *mean* strain leads to a *mean* spin and that spin fluctuations are a higher-order concern. That is reasonable because in the mean-field approximation the strain fluctuations are ignored. These in themselves lead to substantial spin fluctuations, as will be shown later on. The issue is really what exactly is meant by the mean-field approximation. It was seen in the previous chapter that for non-frictional contacts the fabric fluctuations average out. In that case the mean strain approximation is equivalent to a mean fabric approximation. However, in the current configuration correlations between the components of the structural sums emerge and these affect the evaluation of the stress increment.

If the inverse of $\varepsilon_{ijk}\varepsilon_{\ell mn}\sum_{\nu=1}^{N^\mu} K_{j\ell}^{\mu\nu} x_k^{\mu\nu} c_m^{\mu\nu}$ exists, the spin can be expressed in (and is proportional to) the mean strain. In other words

$\eta_n^\mu = \Lambda_{n\ell m}^\mu \bar{e}_{\ell m}$ and the spin fluctuations $\eta_n^\mu - \bar{\eta}_n = \left(\Lambda_{n\ell m}^\mu - \bar{\Lambda}_{n\ell m}\right)\bar{e}_{\ell m}$

If the mean stress increment is calculated the spin needs to be inserted according to the evaluation of the contact forces

$$\begin{aligned}\bar{\sigma}_{ij} &= \frac{1}{2V}\sum_\mu\sum_{\nu=1}^{N^\mu} f_i^{\mu\nu} c_j^{\mu\nu} \\ &= \frac{1}{2V}\sum_\mu\sum_{\nu=1}^{N^\mu} K_{ik}^{\mu\nu}\left[u_k^\nu - u_k^\mu - \varepsilon_{k\ell m}\left(x_\ell^{\mu\nu}\omega_m^\mu - x_\ell^{\nu\mu}\omega_m^\nu\right)\right]c_j^{\mu\nu}\end{aligned}$$

The mean-field estimates follows

$$\bar{\sigma}_{ij} = \frac{1}{2V}\sum_{\mu}\sum_{\nu=1}^{N^\mu} K_{ik}^{\mu\nu}\left(\bar{e}_{k\ell}c_\ell^{\mu\nu} - \varepsilon_{k\ell m}c_\ell^{\mu\nu}\Lambda_{mpq}^{\mu}\bar{e}_{pq}\right)c_j^{\mu\nu}$$

$$= \frac{1}{2V}\sum_{\mu}\sum_{\nu=1}^{N^\mu} K_{ik}^{\mu\nu}\left(\frac{1}{2}\left(\delta_{kp}\delta_{\ell q} + \delta_{kq}\delta_{\ell p}\right)c_\ell^{\mu\nu} - \varepsilon_{k\ell m}c_\ell^{\mu\nu}\Lambda_{mpq}^{\mu}\right)c_j^{\mu\nu}\bar{e}_{pq} = X_{ijpq}^{(mf)}\bar{e}_{pq}$$

Depending then on one's view of what the mean-field approximation represents either $\bar{\mathbf{\Lambda}}$ or $\mathbf{\Lambda}^\mu$ may be used. If the latter is employed a correlation between the components of the structural sums becomes manifest.

These formulas acquire more transparency when examples are investigated. As a simple example, consider a problem in two dimensions with an interaction that is 'diagonal' and the same for all contacts. In addition it has a purely normal and purely tangential form. Particles are discs, therefore $\mathbf{n}^{\mu\nu}$ and $\mathbf{c}^{\mu\nu}$ are aligned. Defining the tangential contact unit vector $\bar{\mathbf{n}}^{\mu\nu}$ (that is, such that $\bar{\mathbf{n}}^{\mu\nu} \perp \mathbf{n}^{\mu\nu}$) the interaction tensor takes the form

$$K_{ij}^{\mu\nu} = k_\perp n_i^{\mu\nu}n_j^{\mu\nu} + k_{//}\bar{n}_i^{\mu\nu}\bar{n}_j^{\mu\nu}$$

If the tangential interactive strength equals the normal strength, the interactive tensor becomes a Kronecker delta. This follows from the algebra of the unit vectors $n_i^{\mu\nu}n_j^{\mu\nu} + \bar{n}_i^{\mu\nu}\bar{n}_j^{\mu\nu} = \delta_{ij}$.

Intriguingly, the problem depends entirely on structural sums of the type

$$\sum_{\nu=1}^{N^\mu} K_{j\ell}^{\mu\nu}x_k^{\mu\nu}c_m^{\mu\nu} \text{ and } \sum_{\nu=1}^{N^\mu} K_{j\ell}^{\mu\nu}c_k^{\mu\nu}c_m^{\mu\nu}$$

For the sake of transparency of the result it is assumed that the particles are more or equal in size, so that

$$\sum_{\nu=1}^{N^\mu} K_{j\ell}^{\mu\nu}x_k^{\mu\nu}c_m^{\mu\nu} \simeq \frac{1}{2}\sum_{\nu=1}^{N^\mu} K_{j\ell}^{\mu\nu}c_k^{\mu\nu}c_m^{\mu\nu} \simeq \frac{1}{2}\bar{c}^2\sum_{\nu=1}^{N^\mu} K_{j\ell}^{\mu\nu}n_k^{\mu\nu}n_m^{\mu\nu}$$

The sum is replaced by an integral, weighed by the contact fabric function $p^\mu(\varphi)$; the latter is — as was done previously — approximated by a quadratic function in the unit normals. All the summations are easily done. In the calculation — without loss of generality — the coordinate frame is chosen in such a way that the mean fabric function is diagonal. The mean-field stiffness (including the correlation due to spin fluctuations) turns out to be

$$\frac{16v}{\pi c^2} X_{1111}^{(mf)} = \overline{p}_{11}\left(5k_{\perp} + k_{//}\right) + \overline{p}_{22}\left(k_{\perp} + k_{//}\right) - 2k_{//}\overline{\left[\frac{\left(p_{12}^{\bullet}\right)^2}{p_{11}^{\bullet} + p_{22}^{\bullet}}\right]};$$

$$\frac{16v}{\pi c^2} X_{1122}^{(mf)} = \frac{16v}{\pi c^2} X_{2211}^{(mf)} = \left(k_{\perp} - k_{//}\right)\left(\overline{p}_{11} + \overline{p}_{22}\right) + 2k_{//}\overline{\left[\frac{\left(p_{12}^{\bullet}\right)^2}{p_{11}^{\bullet} + p_{22}^{\bullet}}\right]};$$

$$\frac{16v}{\pi c^2} X_{2222}^{(mf)} = \overline{p}_{11}\left(k_{\perp} + k_{//}\right) + \overline{p}_{22}\left(5k_{\perp} + k_{//}\right) - 2k_{//}\overline{\left[\frac{\left(p_{12}^{\bullet}\right)^2}{p_{11}^{\bullet} + p_{22}^{\bullet}}\right]};$$

$$\frac{16v}{\pi c^2} X_{1212}^{(mf)} = \frac{16v}{\pi c^2} X_{1221}^{(mf)} = \frac{16v}{\pi c^2} X_{2112}^{(mf)} = \frac{16v}{\pi c^2} X_{2121}^{(mf)}$$

$$= 2k_{\perp}\left(\overline{p}_{11} + \overline{p}_{22}\right) + k_{//}\overline{\left[\frac{\left(p_{11}^{\bullet}\right)^2 + 6p_{11}^{\bullet}p_{22}^{\bullet} + \left(p_{22}^{\bullet}\right)^2}{p_{11}^{\bullet} + p_{22}^{\bullet}}\right]}$$

The result contains various interesting aspects. The system is obviously elastic and the stiffness tensor has the appropriate symmetries. The stress is symmetric, as it should be.

The dependence on the fabric *fluctuations* arises — as mentioned — due to the spin fluctuations. At this stage they should not be taken too seriously, because there are substantial spin fluctuations that are associated with strain fluctuations (these will be estimated below). It is useful though to inspect the order of magnitude of these terms. The largest is $2k_{//}\overline{\left[\left(p_{12}^{\bullet}\right)^2 / \left(p_{11}^{\bullet} + p_{22}^{\bullet}\right)\right]}$. Calculating this contribution, from the simulation data used in Chapter 5, Section 5.1, and comparing it to $\overline{p}_{11} + \overline{p}_{22}$ produces a number in the order of magnitude of $0.1k_{//}$. The other fluctuational contributions are an order smaller than that, a few percent maximum.

Note that all the fluctuational terms in this mean-field calculation are proportional to the tangential spring constant.

A further intriguing possibility introduced by the aspect of the tangential interparticle interaction is the form taken by the off-diagonal moduli. Disregarding effects of fluctuations for the moment, these are proportional to $(k_{\perp} - k_{//})$, implying that they become negative when the tangential interaction is stiffer than the normal interaction. The Poisson ratio would then be negative, which for instance means that when a material is uniaxially compressed it becomes thinner, rather than wider. For contact interactions this will by and large not take place (as generally $k_{\perp} > k_{//}$), but it is possible to create artificial particles with specially manufactured interactive features that do have such properties. Materials consisting of such particles, which possess a negative Poisson ratio are called *auxetic materials* (Gk αυξη — growth or increase), see [Lakes, 1987], [Koenders, 2009] for $k_{\perp} < k_{//}$. They are an example of *meta-materials*, a class of substances with counterintuitive properties. Auxetic foams, that have an internal lever-like structure are non-granular examples of these materials. These are quite easily manufactured and have found many applications.

The effects of strain fluctuations will be assessed against the mean-field estimates produced here.

7.4 Heterogeneity with tangential interactions

The analysis for fabric fluctuations for the case when there are tangential interactions is similar to the one developed for the normal interactive case in the Chapter 6. A refinement to handle the spins needs to be implemented.

The two sets of equations that rule static force and moment incremental equilibrium are

$$\sum_{\nu=1}^{N^\mu} K_{ij}^{\mu\nu}\left[e_{j\ell}^{\mu} c_{\ell}^{\mu\nu} + \tfrac{1}{2}\left.\frac{\partial^2 u_j}{\partial x_k \partial x_\ell}\right|^{\mu} c_k^{\mu\nu} c_\ell^{\mu\nu} - \varepsilon_{jk\ell}\left(c_k^{\mu\nu}\eta_\ell^{\mu} + \tfrac{1}{2}\left.\frac{\partial \eta_\ell}{\partial x_m}\right|^{\mu} c_k^{\mu\nu} c_m^{\mu\nu}\right)\right] = 0$$

$$\varepsilon_{ijk}\sum_{\nu=1}^{N^\mu} K_{j\ell}^{\mu\nu} x_k^{\mu\nu}\left[e_{\ell m}^{\mu} c_m^{\mu\nu} + \tfrac{1}{2}\left.\frac{\partial^2 u_\ell}{\partial x_k \partial x_m}\right|^{\mu} c_k^{\mu\nu} c_m^{\mu\nu} \right.$$

$$\left. -\varepsilon_{\ell mn}\left(\eta_n^\mu c_m^{\mu\nu} + \tfrac{1}{2}\left.\frac{\partial \eta_n}{\partial x_p}\right|^{\mu} c_m^{\mu\nu} c_p^{\mu\nu} \right)\right] = 0$$

The approximation $\mathbf{x}^{\mu\nu} = \tfrac{1}{2}\mathbf{c}^{\mu\nu}$ is again made and a shorthand notation is introduced (see also List of Symbols and Notations, Section 2), which makes the notation tremendously more compact: the structural sums are called $\mathbf{A}$, so

$$A_{ijk}^{\bullet} = \sum_{\nu=1}^{N^\bullet} K_{ij}^{\bullet\nu} c_k^{\bullet\nu};\; A_{ijk\ell}^{\bullet} = \sum_{\nu=1}^{N^\bullet} K_{ij}^{\bullet\nu} c_k^{\bullet\nu} c_\ell^{\bullet\nu};\text{ etc}$$

Averages and fluctuations are now employed. The latter are denoted by a prime; the superscript, denoting the particle number, may be omitted; this will not lead to confusion. The first equation up to first order in the fluctuations is

$$A'_{ij\ell}\overline{e}_{j\ell} + \frac{1}{2}\overline{A}_{ijk\ell}\frac{\partial^2 u'_j}{\partial x_k \partial x_\ell} - \varepsilon_{jk\ell}A'_{ijk}\overline{\eta}_\ell - \frac{1}{2}\varepsilon_{jk\ell}\overline{A}_{ijkm}\frac{\partial \eta'_\ell}{\partial x_m} = 0$$

Similarly, the second equation takes the form

$$\varepsilon_{ijk}\left(A'_{jlkm}\overline{e}_{\ell m} + \overline{A}_{jlkm}e'_{\ell m} - \varepsilon_{\ell mn}A'_{jlkm}\overline{\eta}_n - \varepsilon_{\ell mn}\overline{A}_{jlkm}\eta'_n \right) = 0$$

From this equation the spin fluctuation may be determined. To that end the inverse of $\Xi_{in} = \varepsilon_{ijk}\varepsilon_{\ell mn}\overline{A}_{jlkm}$ must be determined. This is often quite a simple object. Taking the interaction, for example, as before $K_{ij}^{\mu\nu} = k_\perp n_i^{\mu\nu} n_j^{\mu\nu} + k_{//} m_i^{\mu\nu} m_j^{\mu\nu}$, then in two dimensions the only component that is non-zero is $\Xi_{33} = \pi k_{//}\left(\overline{p}_{11} + \overline{p}_{22}\right)$. It simply counts the number of contacts that have a non-zero tangential interaction.

The spin fluctuation is

$$\eta'_\ell = \left(\Xi\right)_{\ell i}^{-1}\left(\varepsilon_{ijk}A'_{jakb}\overline{e}_{ab} + \varepsilon_{ijk}\overline{A}_{jakb}e'_{ab} - \varepsilon_{ijk}\varepsilon_{abn}A'_{jakb}\overline{\eta}_n \right) \quad (7.1)$$

The derivative is required to use in the sum of force increments balance, which is simply

$$\frac{\partial \eta'_\ell}{\partial x_m} = (\Xi)^{-1}_{\ell i}\left(\varepsilon_{ijk}\frac{\partial A'_{jakb}}{\partial x_m}\overline{e}_{ab} + \varepsilon_{ijk}\overline{A}_{jakb}\frac{\partial e'_{ab}}{\partial x_m} - \varepsilon_{ijk}\varepsilon_{abn}\frac{\partial A'_{jakb}}{\partial x_m}\overline{\eta}_n\right)$$

Finally, the odd structural sum is treated as before to give a spatial derivative, which takes place in the same manner as was done for the frictionless case, except that the average interaction now depends on the direction. In order to make that clear, an extra superscript is added to the average bar. Thus, $\overline{\mathbf{K}}^{\mu\nu}$ is the mean value of the interaction in the direction of $\mathbf{n}^{\mu\nu}$. The starting point is

$$\begin{aligned}
&\left(\frac{\partial \sum_{\kappa=1}^{N^\bullet} K_{ij}^{\bullet\kappa} c_k^{\bullet\kappa} c_\ell^{\bullet\kappa}}{\partial x_p}\right)^{\mu} \\
&\simeq \frac{2}{N_V \overline{c}^2}\sum_{\nu=1}^{N^\mu} c_p^{\mu\nu}\sum_{\kappa=1}^{N^\nu} K_{ij}^{\nu\kappa} c_k^{\nu\kappa} c_\ell^{\nu\kappa} \\
&= \frac{2}{N_V \overline{c}^2}\sum_{\nu=1}^{N^\mu}\left(K_{ij}^{\nu\mu} c_k^{\nu\mu} c_\ell^{\nu\mu} c_p^{\mu\nu} + c_p^{\mu\nu}\sum_{\kappa\neq\mu}^{N^\nu} K_{ij}^{\nu\kappa} c_k^{\nu\kappa} c_\ell^{\nu\kappa}\right) \\
&\simeq \frac{2}{N_V \overline{c}^2}\sum_{\nu=1}^{N^\mu}\left(K_{ij}^{\nu\mu} c_k^{\nu\mu} c_\ell^{\nu\mu} c_p^{\mu\nu} + c_p^{\mu\nu}\sum_{\kappa=1}^{N^\nu} \overline{K}_{ij}^{\nu\kappa} n_j^{\nu\kappa} c_k^{\nu\kappa} c_\ell^{\nu\kappa} - \overline{K}_i^{\nu\mu} c_k^{\nu\mu} c_\ell^{\nu\mu} c_p^{\mu\nu}\right) \\
&\simeq \frac{2}{N_V \overline{c}^2}\sum_{\nu=1}^{N^\mu}\left(K_{ij}^{\nu\mu} - \overline{K}_{ij}^{\nu\mu}\right) c_k^{\nu\mu} c_\ell^{\nu\mu} c_p^{\mu\nu}
\end{aligned}$$

And

$$\left(\frac{\partial \sum_{\kappa=1}^{N^\bullet} K_{ij}^{\bullet\kappa} c_k^{\bullet\kappa} c_\ell^{\bullet\kappa}}{\partial x_\ell}\right)^{\mu} \simeq \frac{2}{N_V}\sum_{\nu=1}^{N^\mu}\left(K_{ij}^{\nu\mu} - \overline{K}_{ij}^{\nu\mu}\right) c_k^{\nu\mu}$$

It follows that the translation equilibrium equation takes the approximate form

$$
A'_{ij\ell}\overline{e}_{j\ell}+\frac{1}{2}\overline{A}_{ijk\ell}\frac{\partial^2 u'_j}{\partial x_k\partial x_\ell}-\varepsilon_{jk\ell}A'_{ijk}\overline{\eta}_\ell-\frac{1}{2}\varepsilon_{jk\ell}\overline{A}_{ijkm}\frac{\partial\eta'_\ell}{\partial x_m}=0\rightarrow
$$

$$
\frac{N_V}{2}\frac{\partial A'_{ij\ell m}}{\partial x_m}\left(\overline{e}_{j\ell}-\varepsilon_{j\ell a}\overline{\eta}_a\right)+\frac{1}{2}\overline{A}_{ijk\ell}\frac{\partial^2 u'_j}{\partial x_k\partial x_\ell}
$$

$$
-\frac{1}{2}\varepsilon_{jk\ell}\overline{A}_{ijkm}\left(\left(\Xi\right)^{-1}_{\ell p}\left(\varepsilon_{pqr}\frac{\partial A'_{qarb}}{\partial x_m}\overline{e}_{ab}+\varepsilon_{pqr}\overline{A}_{qarb}\frac{\partial e'_{ab}}{\partial x_m}-\varepsilon_{pqr}\varepsilon_{abn}\frac{\partial A'_{qarb}}{\partial x_m}\overline{\eta}_n\right)\right)=0
$$

Making use of the definition of the strain and rearranging the subscripts shows that the structure of the equilibrium equation is — just as in the case of frictionless contacts — similar to the one obtained in the continuum treatment of heterogeneity.

$$
\begin{aligned}
&\left[\overline{A}_{ijk\ell}-\tfrac{1}{2}\varepsilon_{fgh}\varepsilon_{pqr}\left(\Xi\right)^{-1}_{hp}\overline{A}_{ifgk}\left(\overline{A}_{qjr\ell}+\overline{A}_{q\ell rj}\right)\right]\frac{\partial^2 u'_j}{\partial x_k\partial x_\ell}\\
&+N_V\frac{\partial A'_{iabm}}{\partial x_m}\left(\overline{e}_{ab}-\varepsilon_{abn}\overline{\eta}_n\right)\\
&-\varepsilon_{jk\ell}\varepsilon_{pqr}\left(\Xi\right)^{-1}_{\ell p}\overline{A}_{ijkm}\frac{\partial A'_{qarb}}{\partial x_m}\left(\overline{e}_{ab}-\varepsilon_{abn}\overline{\eta}_n\right)=0
\end{aligned}
\tag{7.2}
$$

Clearly, the same procedure for solving them may be employed as the one that was used for the frictionless case.

7.5 Solution of the strain fluctuations for a two-dimensional isotropic medium

For a two-dimensional isotropic medium the first thing to note is that the terms proportional to $\left(\Xi\right)^{-1}$ in front of the double gradient vanish. In that case the inverse of the acoustic tensor is with $\overline{p}=\overline{p}_{11}=\overline{p}_{22}$

$$
P^{-1}_{ij}=\frac{p_1}{k^4}k_ik_j+\frac{p_0}{k^2}\delta_{ij}
$$

$$
\text{with }p_0=\frac{4}{\pi\overline{pc}^2\left(k_\perp+3k_{//}\right)};\ p_1=-\frac{8\left(k_\perp-k_{//}\right)}{\pi\overline{pc}^2\left(k_\perp+3k_{//}\right)\left(3k_\perp+k_{//}\right)}
\tag{7.3}
$$

The solution for the displacement fluctuation is entirely analogous to the frictionless case

$$u'_a(\mathbf{x}) = -\frac{\bar{a}}{8} Z'_{ij}(0)\, p_1 \left[S_1\left(\frac{x}{\bar{a}}\right) + S_3\left(\frac{x}{\bar{a}}\right) \right] \left(m_i \delta_{aj} + m_j \delta_{ai} + m_a \delta_{ij} \right)$$
$$\frac{\bar{a}}{2} Z'_{ij}(0)\, p_1 S_3\left(\frac{x}{\bar{a}}\right) m_i m_j m_a - \frac{\bar{a}}{2} Z'_{ij}(0)\, p_0 m_j \delta_{ai} S_1\left(\frac{x}{\bar{a}}\right) \tag{7.4}$$

With $\mathbf{Z}(0)$ the source term of the differential equation (7.2)

$$Z'_{ij}(0) = N_V A'_{icdj}(0)\left(\bar{e}_{cd} - \varepsilon_{cdn}\bar{\eta}_n\right) - \varepsilon_{ef\ell}\varepsilon_{pqr}\left(\Xi\right)^{-1}_{\ell p} \bar{A}_{iefj} A'_{qcdr}(0)\left(\bar{e}_{cd} - \varepsilon_{cdn}\bar{\eta}_n\right)$$

The extra stress (over and above the mean strain stress) due to the fluctuations is

$$\bar{\sigma}'_{st} = \overline{\sum_{\nu=1}^{N^\bullet} K'^{\bullet\nu}_{su} c^{\bullet\nu}_t \left(u'^{\bullet\nu}_u - \varepsilon_{uk\ell}\left(c^{\bullet\nu}_k \eta'^{\bullet}_\ell + \frac{1}{2}\frac{\partial \eta'_\ell}{\partial x_m}\bigg|^{\bullet} c^{\bullet\nu}_k c^{\bullet\nu}_m \right)\right)}$$

The term proportional to the spin gradient comprises an odd fluctuating structural sum and a fluctuating quantity, so it is of third order in the fluctuations and will average to zero. The fluctuating spin is derived from expression (7.1)

$$\eta'_\ell = \left(\Xi\right)^{-1}_{\ell i}\left(\varepsilon_{ijk} A'_{jakb}\bar{e}_{ab} + \varepsilon_{ijk}\bar{A}_{jakb} e'_{ab} - \varepsilon_{ijk}\varepsilon_{abn} A'_{jakb}\bar{\eta}_n \right)$$

This, in turn, requires the strain fluctuation, which is obtained by differentiating expression (7.4). It is noted that

$$\left.\frac{\partial S_1\left(\frac{x}{\bar{a}}\right)}{\partial x}\right|_{x=0} = \frac{1}{\bar{a}}; \quad \left.\frac{\partial S_3\left(\frac{x}{\bar{a}}\right)}{\partial x}\right|_{x=0} = 0$$

Altogether, it is seen that the evaluation of the effect of heterogeneity is rather more complex in the case of tangential interactions, than for the frictionless case. A symbolic manipulation program is employed. The results are equally complicated due to the rather large number of symbols that is involved. However, a good impression of the result may be obtained by studying special cases.

7.6 Considerations for an isotropic contact distribution in two special cases

The two special cases that will be reported below are: (1) $k_{//} = k_{\perp}$ and (2) $k_{//} = \frac{1}{2} k_{\perp}$.

The same approach to the evaluation of the correlations is followed as in the development of the results of frictionless particles. So, the contact point distribution is set to a quadratic form

$$p^{\mu}(\varphi) = p_{ij}^{\mu} n_i(\varphi) n_j(\varphi) = \bar{p} + p_{ij}^{\prime\mu} n_i(\varphi) n_j(\varphi)$$

Using this, the fluctuation in the spin is evaluated

$$\eta' = \frac{p'_{11} + p'_{22}}{4\bar{p}} \bar{\eta} - \frac{p'_{11} - p'_{22}}{4\bar{p}} e'_{12} + \frac{p'_{12}}{4\bar{p}} (e'_{11} - e'_{22})$$

The way is now open to calculate the stress contribution due to fluctuations in the contact point distribution up to quadratic order. A slightly more restrictive approach is taken in that it will be assumed that the fluctuations are isotropically distributed. So, any fluctuation can be described as

$$p'_{ij} = P' \delta_{ij} + \tilde{p}'_{ij}$$

The deviatoric part can be written as a rotated trace-free tensor. Call the rotation $\mathbf{Q}(\alpha)$, then

$$\tilde{p}'_{ij} = Q_{ik}(\alpha) Q_{j\ell}(\alpha) d_{k\ell}, \text{ with } \mathbf{d} \to \begin{pmatrix} \delta p' & 0 \\ 0 & -\delta p' \end{pmatrix}$$

Averaging over all possible angles α, gives the following

$$\overline{(\tilde{p}'_{11})^2} = \overline{(\tilde{p}'_{12})^2} = \overline{(\tilde{p}'_{22})^2} = \frac{1}{2}\overline{(\delta p')^2}; \overline{(\tilde{p}'_{11}\tilde{p}'_{22})} = -\frac{1}{2}\overline{(\delta p')^2};$$

all other correlations zero.

The stress should of course be symmetric and this fact can be used to express the mean spin $\overline{\eta}$. In a perfectly isotropic environment $\overline{\eta}=0$, so in order to illustrate how $\overline{\eta}$ becomes manifest, the difference between $\overline{(\tilde{p}'_{11})^2}$ and $\overline{(\tilde{p}'_{22})^2}$ is allowed to retain a small value $\overline{(\varepsilon')^2}$.

Case (1): $k_{//}=k_{\perp}$

In this case $p_1=0$ and as a result S_3 does not enter the calculation. Up to first order in $\overline{(\varepsilon')^2}$ the mean spin is derived from the imposition of stress symmetry

$$\overline{\eta}\simeq\frac{2\overline{(\varepsilon')^2}\left(2N_V\frac{\hat{a}}{\overline{c}}S_1-\frac{\hat{a}}{\overline{c}}S_1+2\right)}{N_V\frac{\hat{a}}{\overline{c}}\left(4\overline{(\delta p')^2}+16\overline{(P')^2}+5\overline{(\varepsilon')^2}\right)S_1-4\left(\overline{(P')^2}+\overline{(\varepsilon')^2}\right)\left(\frac{\hat{a}}{\overline{c}}S_1-2\right)}\overline{e}_{12}$$

Although the result is a little contrived because of the isotropy of the problem, it is seen that the mean spin is proportional to the mean shear stress and that if absolute isotropy is required (that is, $\overline{(\varepsilon')^2}\to 0$) the mean spin vanishes. The relative shear modulus correction turns out to be simply

$$\frac{\mu-\mu^{(mf)}}{\mu^{(mf)}}=-\frac{2N_V\frac{\hat{a}}{\overline{c}}\left(4\overline{(P')^2}+\overline{(\delta p')^2}\right)S_1+\overline{(\delta p')^2}\left(2-\frac{\hat{a}}{\overline{c}}S_1\right)}{16\overline{p}^2}$$

The leading term is the one that contains the variability in the mean number of contacts $\overline{(P')^2}$; it is observed that for the not unreasonable value of $(\hat{a}/c)S_1=0.5$ and $N_V=6$ the correction is of the order of $-\frac{3}{2}\overline{(P')^2}/\overline{p}^2$, which — using the same estimate for the variability as in the previous chapter — is about half the value of the frictionless case. Thus, the packed bed with a fully frictional interaction is far less sensitive to fluctuations than the same bed with a frictionless interaction.

The correction to the λ Lamé constant, here scaled to the mean-field shear modulus, is

$$\frac{\lambda - \lambda^{(mf)}}{\mu^{(mf)}} = -\frac{\left(2 - \frac{\hat{a}}{\bar{c}} S_1\right)\overline{(\delta p')^2}}{32\bar{p}^2}$$

This is a small correction, though note that it takes the material into the auxetic range.

Case (2): $k_{//} = \frac{1}{2}k_{\perp}$

When $k_{//} \neq k_{\perp}$ S_3 enters the calculation. This is immediately clear when the mean spin is calculated up to first order in $\overline{(\varepsilon')^2}$

$$\bar{\eta} \simeq \frac{\overline{(\varepsilon')^2}}{4}\bar{e}_{12} \times$$

$$\frac{N_V \frac{\hat{a}}{\bar{c}}(32S_1 - 3S_3) - 7\left(\frac{\hat{a}}{\bar{c}}S_1 - 5\right)}{N_V \frac{\hat{a}}{\bar{c}}(2\left(3\overline{(\delta p')^2} + 14\overline{(P')^2}\right)S_1 - S_3\overline{(\delta p')^2}) - 14\overline{(P')^2}\left(\frac{2\hat{a}}{\bar{c}}S_1 - 5\right)}$$

The shear modulus correction takes the form

$$\frac{\mu - \mu^{(mf)}}{\mu^{(mf)}} = -\frac{6N_V \frac{\hat{a}}{\bar{c}}\left[9\left(4\overline{(P')^2} + \overline{(\delta p')^2}\right)S_1\right] + 7\overline{(\delta p')^2}\left(5 - 2\frac{\hat{a}}{\bar{c}}S_1\right)}{420\bar{p}^2}$$

$$+\frac{N_V \frac{\hat{a}}{\bar{c}}\left[S_3\left(2\overline{(P')^2} + \overline{(\delta p')^2}\right)\right]}{35\bar{p}^2}$$

The leading term is now of the order of $-1.5\overline{(P')^2}/\bar{p}^2$, so it is more sensitive to fluctuations than the previous case, but not as susceptible as the frictionless case.

It is also found that

$$\frac{\lambda-\lambda^{(mf)}}{\overline{\mu}}=-\frac{N_V\frac{\hat{a}}{\overline{c}}\left[\left(\overline{52\overline{(P')}^2+21\overline{(\delta p')}^2}\right)S_1-2S_3\left(\overline{18\overline{(P')}^2-\overline{(\delta p')}^2}\right)\right]}{70\overline{p}^2}$$

$$-\frac{\overline{(\delta p')}^2\left(5-2\frac{\hat{a}}{\overline{c}}S_1\right)}{120\overline{p}^2}$$

The leading term is of the order $-0.37\overline{(P')}^2/\overline{p}^2$, so this is also small and, again, makes the modulus smaller.

7.7 Anisotropic calculation

A slightly different approach is taken to calculate the effective moduli for an anisotropic medium due to heterogeneity. The purpose of this calculation is basically to show the effects of intrinsic directional properties. The mean field values have already been evaluated in Section 7.3. In order to ascertain the effects of anisotropy in two dimensions the starting point is again the two equilibrium equations in which the spins are decomposed into an average and fluctuations

$$\sum_{\nu=1}^{N^\mu}K_{ij}^{\mu\nu}\left[e_{j\ell}^{\mu}c_\ell^{\mu\nu}+\frac{1}{2}\frac{\partial^2u_j}{\partial x_k\partial x_\ell}\bigg|^\mu c_k^{\mu\nu}c_\ell^{\mu\nu}-\varepsilon_{jk\ell}\left(\frac{1}{2}c_k^{\mu\nu}\left(\eta_\ell'^{\mu}+\eta_\ell'^{\nu}\right)+c_k^{\mu\nu}\overline{\eta}_\ell\right)\right]=0$$

$$\varepsilon_{ijk}\sum_{\nu=1}^{N^\mu}K_{j\ell}^{\mu\nu}x_k^{\mu\nu}\left[e_{\ell m}^{\mu}c_m^{\mu\nu}+\frac{1}{2}\frac{\partial^2u_\ell}{\partial x_k\partial x_m}\bigg|^\mu c_k^{\mu\nu}c_m^{\mu\nu}\right.$$

$$\left.-\varepsilon_{\ell mn}\left(\eta_n^{\mu}c_m^{\mu\nu}+\frac{1}{2}\frac{\partial\eta_n}{\partial x_p}\bigg|^\mu c_m^{\mu\nu}c_p^{\mu\nu}\right)\right]=0$$

Using the shorthand notation introduced for the structural sums as before, the latter equation takes the form

$$\varepsilon_{ijk}\left(A^{\mu}_{j\ell km}e^{\mu}_{\ell m}+\frac{1}{2}A^{\mu}_{j\ell knm}\left.\frac{\partial^2 u_\ell}{\partial x_n \partial x_m}\right|^{\mu}-\varepsilon_{\ell mn}A^{\mu}_{j\ell km}\left(\frac{1}{2}\eta'^{\mu}_n+\overline{\eta}_n\right)\right)$$
$$-\frac{1}{2}\varepsilon_{\ell mn}\sum_{\nu=1}^{N^{\mu}}K^{\mu\nu}_{j\ell}x^{\mu\nu}_k c^{\mu\nu}_m \eta'^{\nu}_n=0$$

An approximation is made: the sum over the neighbouring spins and the double fluctuations involving the second displacement gradient are both neglected.

Solve for η'_3 in two dimensions for the interaction $k_{//}=k_{\perp}$

$$\eta'=-2\overline{\eta}+\frac{\pi dp_{12}}{N^{\mu}_c}\left(\overline{e}_{11}-\overline{e}_{22}\right)-\frac{\pi\left(\overline{p}_{11}-\overline{p}_{22}+dp_{11}-dp_{22}\right)}{N^{\mu}_c}\overline{e}_{12}$$
$$+\frac{\pi}{N^{\mu}_c}\left(dp_{12}\left(de_{11}-de_{22}\right)-\left(\overline{p}_{11}-\overline{p}_{22}+dp_{11}-dp_{22}\right)de_{12}\right)$$

The force equilibrium equation is written up to first order in the fluctuations

$$\sum_{\nu=1}^{N^{\mu}}K^{\mu\nu}_{ij}c^{\mu\nu}_{\ell}\left(\overline{e}_{j\ell}-\varepsilon_{j\ell k}\overline{\eta}_k\right)+\frac{1}{2}\overline{\sum_{\nu=1}^{N^{\bullet}}K^{\bullet\nu}_{ij}c^{\bullet\nu}_k c^{\bullet\nu}_{\ell}}\left.\frac{\partial^2 u_j}{\partial x_k \partial x_\ell}\right|^{\mu}=0$$

This equation may again be solved by replacing the fluctuating structural sum by a derivative of a higher order structural sum and take the fluctuating part. Then the whole system is as before, with the difference that the average second order structural sum is now anisotropic. In other words

$$N_V\frac{\partial A'_{ij\ell k}}{\partial x_k}\left(\overline{e}_{j\ell}-\varepsilon_{j\ell k}\overline{\eta}_k\right)+\overline{A}_{ijk\ell}\frac{\partial^2 u'_j}{\partial x_k \partial x_\ell}=0$$

The solution is again found by Fourier transform

$$u'_a(\mathbf{x})=\frac{iN_V}{(2\pi)^2}\int d_2 k e^{i\mathbf{k}.\mathbf{x}}P^{-1}_{ai}k_k\int d_2 y A'_{ij\ell k}(\mathbf{y})e^{-i\mathbf{k}.\mathbf{y}}\left(\overline{e}_{j\ell}-\varepsilon_{j\ell k}\overline{\eta}_k\right),$$

where $P_{ij}=\overline{A}_{ijk\ell}k_k k_\ell$.

In the case of $k_{//} = k_{\perp}$, this acoustic tensor is diagonal and takes the following form in two dimensions

$$\mathbf{P} = \frac{\pi \overline{c}^2}{4} k_{\perp} \left[\left(3\overline{p}_{11} + \overline{p}_{22} \right) k_1^2 + \left(\overline{p}_{11} + 3\overline{p}_{22} \right) k_2^2 \right] \boldsymbol{\delta}$$

The inverse is easily obtained

$$\mathbf{P}^{-1} = \frac{4}{\pi \overline{c}^2 k_{\perp} \left[\left(3\overline{p}_{11} + \overline{p}_{22} \right) k_1^2 + \left(\overline{p}_{11} + 3\overline{p}_{22} \right) k_2^2 \right]} \boldsymbol{\delta}$$

The integrals can be attempted for the case that $A'_{ij\ell k}(\mathbf{y}) = A'_{ij\ell k}(0) \exp\left(-y^2 / \widehat{a}^2\right)$. First the inner integral is evaluated

$$\int d_2 y A'_{ij\ell k}(\mathbf{y}) e^{-i\mathbf{k}\cdot\mathbf{y}} = \pi \widehat{a}^2 \exp\left(-\frac{1}{4} \widehat{a}^2 k^2 \right) A'_{ij\ell k}(0)$$

The integral over k gives rise to a confluent hypergeometric function (sometimes called a Kummer function), see Appendix, Section A.6.2. The subsequent integral over ψ may be done numerically, or by a (high-order) series expansion. The problem is *not* getting an answer, the problem is getting an answer that is transparent, insofar as it illuminates the anisotropic character of the outcome. The result will be a expansion in the angle φ in terms of the trigonometric functions $\cos\varphi$ and $\sin\varphi$. From the numerical work it is found that the lowest terms in the expansion contribute to any significant amount only. Therefore, these low-order terms are obtained by searching for the lowest terms of a Fourier expansion. This task is easy enough to perform. Change the integration over the angle ψ to $\alpha = \psi - \varphi$, then

$$u'_a(\mathbf{x}) = \frac{i\hat{a}^2 N_V}{4\pi} \int_0^{2\pi} d\psi \hat{P}_{ai}^{-1}(\psi)\hat{n}_k(\psi)$$

$$\times \int_0^{\infty} dk e^{ikx\cos(\varphi-\psi)} \exp\left(-\frac{1}{4}\hat{a}^2 k^2\right) A'_{ij\ell k}(0)\left(\overline{e}_{j\ell} - \varepsilon_{j\ell k}\overline{\eta}_k\right)$$

$$= \frac{-\hat{a}^2 N_V}{4\pi} \int_0^{2\pi} d\alpha \hat{P}_{ai}^{-1}(\alpha+\varphi)\hat{n}_k(\alpha+\varphi)$$

$$\times \int_0^{\infty} dk \sin\left[kx\cos(\alpha)\right] \exp\left(-\frac{1}{4}\hat{a}^2 k^2\right) A'_{ij\ell k}(0)\left(\overline{e}_{j\ell} - \varepsilon_{j\ell k}\overline{\eta}_k\right)$$

The Fourier coefficients are obtained from

$$\frac{1}{\pi}\int_0^{2\pi} \hat{P}_{ai}^{-1}(\alpha+\varphi)\hat{n}_k(\alpha+\varphi)\begin{pmatrix}\cos(m\varphi)\\ \sin(m\varphi)\end{pmatrix} d\varphi$$

Writing, $\overline{p}_{11} = \overline{p}(1+a_p)$ and $\overline{p}_{22} = \overline{p}(1-a_p)$, the result is for $m=1$, depending on the value of the subscripts

$$a=i;k=1: \frac{2}{\pi k_\perp \overline{c}^2 \overline{p} a_p \sqrt{a_p+2}}\begin{pmatrix}\left(\sqrt{a_p+2}-\sqrt{2-a_p}\right)\cos\alpha\\ \left(\sqrt{2-a_p}-\sqrt{2+a_p}\right)\sin\alpha\end{pmatrix};$$

$$a=i;k=2: \frac{2}{\pi k_\perp \overline{c}^2 \overline{p} a_p \sqrt{2-a_p}}\begin{pmatrix}\left(\sqrt{a_p+2}-\sqrt{2-a_p}\right)\sin\alpha\\ \left(\sqrt{a_p+2}-\sqrt{2-a_p}\right)\cos\alpha\end{pmatrix}$$

The results for even m are all zero and for higher, odd values of m they are substantially smaller than for $m=1$.

The integrals that need to be done are now

$$\int_0^{2\pi} d\alpha \begin{pmatrix}\cos\alpha\\ \sin\alpha\end{pmatrix}\int_0^{\infty} dk \sin\left[kx\cos(\alpha)\right]\exp\left(-\frac{1}{4}\hat{a}^2k^2\right)$$

The integral with $\sin\alpha$ is zero. The integral with $\cos\alpha$ has already been encountered (see Appendix, Section A.6.1)

$$\int_0^{2\pi} d\alpha \cos\alpha \int_0^{\infty} dk \sin\left[kx\cos(\alpha)\right]\exp\left(-\frac{1}{4}\hat{a}^2k^2\right)$$
$$= 2\pi\int_0^{\infty} dk J_1(kx)\exp\left(-\frac{1}{4}\hat{a}^2k^2\right) = \frac{2\pi}{\hat{a}}S_1\left(\frac{x}{\hat{a}}\right)$$

Altogether, the result is

$$u_i'(\mathbf{x}) = -\frac{2\hat{a}N_V S_1\left(\dfrac{x}{\hat{a}}\right)}{\pi k_\perp \overline{pc}^2} C_k(a_p,\varphi) A'_{ij\ell k}(0)\left(\overline{e}_{j\ell} - \varepsilon_{j\ell k}\overline{\eta}_k\right)$$

with

$$C_1(a_p,\varphi) = \frac{\sqrt{a_p+2}-\sqrt{2-a_p}}{2a_p\sqrt{a_p+2}}\cos\varphi;$$
$$C_2(a_p,\varphi) = \frac{\sqrt{a_p+2}-\sqrt{2-a_p}}{2a_p\sqrt{2-a_p}}\sin\varphi$$

In the limit $a_p \to 0$ this reverts back to the isotropic case. The two functions

$$C^{(1)}(a_p) = \frac{\sqrt{a_p+2}-\sqrt{2-a_p}}{2a_p\sqrt{a_p+2}} \text{ and } C^{(2)}(a_p) = \frac{\sqrt{a_p+2}-\sqrt{2-a_p}}{2a_p\sqrt{2-a_p}}$$

are plotted in Fig. 7.1 in the range $-1 < a_p < 1$.

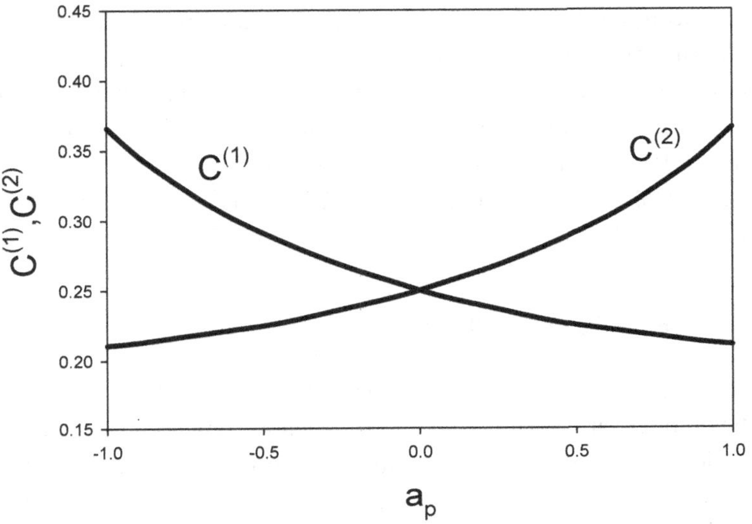

Figure 7.1. The factors $C^{(1)}$ and $C^{(2)}$ as a function of a_p.

The two diagonal stresses are evaluated. As before, set $p'_{11} = P' + p'$ and $p'_{22} = P' - p'$. Here, only the terms that pertain to the isotropic variations in contact point distributions are reported. The reason for this is that this is by far the most significant term and also that in an anisotropic setting it is not directly obvious what the deviatoric distribution of contacts may plausibly look like. Furthermore, a presentational point is that including deviatoric information makes the outcome very opaque due to the large number of symbols involved.

$$\frac{2v}{\pi \bar{c}^2 k_\perp}\bar{\sigma}_{11} = \frac{1}{2}\bar{p}\left(2 + a_p\right)\bar{e}_{11} - 2N_V \frac{\overline{(P')^2}}{\bar{p}}\frac{\bar{c}}{\bar{a}}S_1\left(\frac{\bar{c}}{\bar{a}}\right)C^{(1)}\left(a_p\right)\bar{e}_{11} \tag{7.5}$$

$$\frac{2v}{\pi \bar{c}^2 k_\perp}\bar{\sigma}_{22} = \frac{1}{2}\bar{p}\left(2 - a_p\right)\bar{e}_{22} - 2N_V \frac{\overline{(P')^2}}{\bar{p}}\frac{\bar{c}}{\bar{a}}S_1\left(\frac{\bar{c}}{\bar{a}}\right)C^{(2)}\left(a_p\right)\bar{e}_{22} \tag{7.6}$$

The dependence on the anisotropy in the 22 component is obtained by making the substitution $a_p \rightarrow -a_p$ in the 11 component. From the plot of

$C^{(1,2)}$ it is seen that the weaker stiffness direction is more affected by the effects of anisotropy than the stiffer one.

The higher order terms, that is the contributions that pertain to $m = 3,5,7$, etc., also play a role, though they are small compared to the $m = 1$ case (largely due to the fact that they give rise to higher order $S_{\bullet}$ functions). The approximation presented here is very acceptable and useable.

7.8 A few remarks on the theory

The approximations that have been made to arrive at these results have either been idealisations of the geometry, such as replacing the location of the contact point with half the branch vector, or neglecting double fluctuations or sums over fluctuating quantities of a ring of neighbours. Each of these approximations can be investigated further as refinements, probably at the expense of introducing more parameters. Also note that introducing extra fluctuating terms in the analysis only makes sense when they appear quadratically in the end result. However, the main findings of this chapter relate to the influence of the fabric fluctuations — predominantly the variability in the number of contacts per particle — and the effect of anisotropy on the sensitivity of the assembly to fabric fluctuations. These are the dominant effects that influence the order of magnitude of the outcome as far as the incremental stiffness components are concerned.

It must be emphasised that the calculation on anisotropy is only valid for the simple case $k_{\perp} = k_{\parallel}$, which leads to the very convenient form for the acoustic tensor. Nevertheless, this calculation is a very useful first step for more complicated cases, as will be seen in Chapter 8.

References

Koenders, M.A. (2009) Wave propagation through elastic granular and granular auxetic materials. *Physica Status Solidi (b)* **246**(9) 2083–2088.

Lakes, R.S. (1987) Foam structures with a negative Poisson's ratio *Science*, **235**(4792) 1038–1040.

Chapter 8

Frictional Granular Materials

8.1 The frictional interaction

In this chapter the incremental stress-strain relation for an assembly in which the frictional interaction plays a role. The procedure is similar to the ones outlined in the previous chapters. First a mean-field approximation is explored and then corrections are introduced to account for heterogeneity and strain fluctuations.

The frictional interaction has already been introduced in Section 1.5. It has two states: a sliding state and a sticking state. Those ideas for an incremental contact law are now applied to an assembly. The sliding interaction itself is summarised here first.

The frictional interaction for a sliding contact with direction unit normal $\mathbf{n}$ is treated as follows. The force on the contact is $\mathbf{F}$; the normal force is $F_i n_i$, the tangential force $F_i \bar{n}_i$. A contact displacement $\mathbf{d}$ must be such that the ratio $F_i \bar{n}_i / F_i n_i$ remains constant. So, as the increment in the force is $f_i = K_{ij} d_j$, invariance of the force ratio as the increment is applied requires

$$\frac{(F_i + f_i)\bar{n}_i}{(F_p + f_p)n_p} = \frac{F_i \bar{n}_i}{F_p n_p} \rightarrow f_p n_p F_i \bar{n}_i - f_i \bar{n}_i F_p n_p = 0$$

This is the same result as the one obtained in Section 1.5, without the need for a Taylor expansion. Then, using $f_i = K_{ij} d_j$,

$$K_{ij} d_j \bar{n}_i - \frac{F_i \bar{n}_i K_{pq} d_q n_p}{F_k n_k} = \left(K_{ia} \bar{n}_i - \frac{F_i \bar{n}_i}{F_k n_k} K_{pa} n_p \right) d_a$$

If, for purely tangential deformation, the stiffness is assumed to be zero then the interaction has the form

$$K_{ij} = kn_i n_j + \frac{F_p \bar{n}_p}{F_k n_k} k\bar{n}_i n_j$$

This is providing that the direction of the displacement is such that no elastic unloading is invoked, which must be verified afterwards when the increment has been applied. The tangential to normal force ratio at sliding is called $\mu_s = \left(F_p \bar{n}_p\right) / \left(F_k n_k\right)$. Its numerical value is a material constant, but its sign depends on the choice of coordinate frame.

The frictional interaction brings with it its own idiosyncracies. In order to explore the problems the sticking state is represented in a very simple form as

$$K_{ij} = k\delta_{ij}$$

All manner of refinements and complexities are possible, some of which will be discussed below, but first the simplest form in an assembly with at most one sliding contact per particle is investigated. For this case the sliding contact is the only perturbation and the point of the exercise is to figure out how the frictional disturbance influences the displacement and spin fields in the vicinity of the anomalous contact. At first sight this is quite simple, as all that needs to be done is to introduce the perturbation into the analysis that has been done in the previous chapter and graft it on. There is a complication however and it is this: if there is a sliding contact of the particle pair $(\mu\sigma)$, then there is also a sliding contact as viewed from particle σ, that is the pair $(\sigma\mu)$ (the sliding contact is denoted by σ, while all the other neighbours are, as before, called ν). These two can obviously not be treated independently when assessing the perturbation field. So, instead of a perturbation due to the interactions of one particle, the perturbations due to a *particle pair* have to be accounted for.

8.2 Mean-field estimate

In the following everything that has been discussed in Chapter 7 will be used as an underlying basis for the analysis. Essentially, maximally one sliding contact per particle is considered only and the theory of the previous chapter is modified to accommodate that. The particles are discs; the assembly is two-dimensional. All idealisations and approximations from the previous chapter are used.

The first task is to ascertain the impact of sliding contacts on the mean-field estimate of the incremental stiffness. To that end a stress increment is evaluated while the displacement fluctuations are ignored. What cannot be ignored, however, is the spin fluctuation, which follows from the moment equation of each particle

$$\varepsilon_{ijk}\sum_{\nu=1}^{N^\mu} f_j^{\mu\nu} x_k^{\mu\nu} = 0 \rightarrow \varepsilon_{ijk}\sum_{\nu=1}^{N^\mu} K_{j\ell}^{\mu\nu} x_k^{\mu\nu}\left[\overline{e}_{\ell m} c_m^{\mu\nu} - \varepsilon_{\ell mn}\left(x_m^{\mu\nu}\eta_n^\mu - x_m^{\nu\mu}\eta_n^\nu\right)\right] = 0$$

All contacts stick, except the $(\mu\sigma)$ one, so in the moment equation for particle μ one sticking contact is subtracted from the summation and replaced with a sliding contact. The moment equation then takes the form

$$\begin{aligned}&\varepsilon_{ijk}\sum_{\nu=1}^{N^\mu} k\delta_{j\ell} x_k^{\mu\nu}\left[\overline{e}_{\ell m} c_m^{\mu\nu} - \varepsilon_{\ell mn}\left(x_m^{\mu\nu}\eta_n^\mu - x_m^{\nu\mu}\eta_n^\nu\right)\right]\\ &+\varepsilon_{ijk} k\left(n_j^{\mu\sigma} n_\ell^{\mu\sigma} + \mu^{\mu\sigma}\overline{n}_j^{\mu\sigma} n_\ell^{\mu\sigma} - \delta_{j\ell}\right) x_k^{\mu\sigma}\left[\overline{e}_{\ell m} c_m^{\mu\sigma} - \varepsilon_{\ell mn}\left(x_m^{\mu\sigma}\eta_n^\mu - x_m^{\sigma\mu}\eta_n^\sigma\right)\right] = 0\end{aligned}$$

The equivalent expression for particle σ is

$$\begin{aligned}&\varepsilon_{ijk}\sum_{\varepsilon=1}^{N^\sigma} k\delta_{j\ell} x_k^{\sigma\varepsilon}\left[\overline{e}_{\ell m} c_m^{\sigma\varepsilon} - \varepsilon_{\ell mn}\left(x_m^{\sigma\varepsilon}\eta_n^\sigma - x_m^{\varepsilon\sigma}\eta_n^\varepsilon\right)\right]\\ &+\varepsilon_{ijk} k\left(n_j^{\sigma\mu} n_\ell^{\sigma\mu} + \mu^{\sigma\mu}\overline{n}_j^{\sigma\mu} n_\ell^{\sigma\mu} - \delta_{j\ell}\right) x_k^{\sigma\mu}\left[\overline{e}_{\ell m} c_m^{\sigma\mu} - \varepsilon_{\ell mn}\left(x_m^{\sigma\mu}\eta_n^\sigma - x_m^{\mu\sigma}\eta_n^\mu\right)\right] = 0\end{aligned}$$

The unit normals satisfy $\mathbf{n}^{\mu\sigma} = -\mathbf{n}^{\sigma\mu}$ and the friction coefficient retains its sign when the superscripts are interchanged. Thus, it follows that the second term in the above equations are identical, but only when the surrounding fabric is equal. In order to illuminate the structure of the addition of friction to the analysis, it is for the moment assumed that there are no fabric variations. For the first terms write the spins as $\boldsymbol{\eta}^{\mu} = \overline{\boldsymbol{\eta}} + \boldsymbol{\eta}'^{\mu}$ and $\boldsymbol{\eta}^{\sigma} = \overline{\boldsymbol{\eta}} + \boldsymbol{\eta}'^{\sigma}$, respectively and in the summation over the neighbours assume that the spin fluctuations of the neighbours average to zero. It is seen that the two equations are now the same, other than the term proportional to the spin fluctuations of the central particle. Subtracting the two equations yields the result that the two spin fluctuations $\boldsymbol{\eta}'^{\mu}$ and $\boldsymbol{\eta}'^{\sigma}$ are equal.

It helps the insight to do an explicit calculation of the problem. For reference the case of all-stick is noted; the formulas from Section 7.3 are applied to the present problem. In this case all spin fluctuations are obviously zero and the mean spin is

$$\left(\overline{\eta}_3\right)_{all\ stick} = \frac{\overline{p}_{22} - \overline{p}_{11}}{2\left(\overline{p}_{11} + \overline{p}_{22}\right)}\overline{e}_{12}$$

The stress increment becomes

$$2v\left(\sigma_{11}\right)_{all\ stick} = \frac{\pi}{4}k\overline{c}^2\left(3\overline{p}_{11} + \overline{p}_{22}\right)\overline{e}_{11};$$

$$2v\left(\sigma_{22}\right)_{all\ stick} = \frac{\pi}{4}k\overline{c}^2\left(\overline{p}_{11} + 3\overline{p}_{22}\right)\overline{e}_{22};$$

$$2v\left(\sigma_{12}\right)_{all\ stick} = 2v\left(\sigma_{21}\right)_{all\ stick} = \frac{\pi}{8}k\overline{c}^2\frac{\left(\overline{p}_{11} + 3\overline{p}_{22}\right)\left(3\overline{p}_{11} + \overline{p}_{22}\right)}{\overline{p}_{11} + \overline{p}_{22}}\overline{e}_{12}$$

These formulas are valid when *all* particles in the assembly have no-slip contacts. If, however, some particles possess a slipping contact, then the mean spin may have a different value. For this case the spin fluctuation of a particle with all-sticking contacts is

$$\left(\eta_3'\right)_{all\ stick} = -2\overline{\eta} - \frac{\overline{p}_{11} - \overline{p}_{22}}{2\left(\overline{p}_{11} + \overline{p}_{22}\right)}\overline{e}_{12}$$

The one-particle stress contributions are the same as in the case of all particles sticking. The mean spin simply does not appear in the formulas. The stress is symmetric (as expected, because moment equilibrium has been enforced).

For the case of a particle with one slipping contact $(\mu\sigma)$ the moment equilibrium is expanded as follows

$$
\begin{aligned}
&\varepsilon_{ijk}\sum_{\nu\neq\sigma}^{N^\mu} k\delta_{j\ell}x_k^{\mu\nu}\left[\overline{e}_{\ell m}c_m^{\mu\nu}-\varepsilon_{\ell mn}c_m^{\mu\nu}\left(\overline{\eta}_n+\frac{1}{2}\eta_n'^{\mu}+\frac{1}{2}\eta_n'^{\sigma}\right)\right]+\\
&\varepsilon_{ijk}k\left(n_j^{\mu\sigma}n_\ell^{\mu\sigma}+\mu^{\mu\sigma}\overline{n}_j^{\mu\sigma}n_\ell^{\mu\sigma}\right)x_k^{\mu\sigma}\left[\overline{e}_{\ell m}c_m^{\mu\sigma}-\varepsilon_{\ell mn}\left(x_m^{\mu\sigma}\eta_n^{\mu}-x_m^{\sigma\mu}\eta_n^{\sigma}\right)\right]=0\\
&\rightarrow\varepsilon_{ijk}\sum_{\nu=1}^{N^\mu} k\delta_{j\ell}x_k^{\mu\nu}\left[\overline{e}_{\ell m}c_m^{\mu\nu}-\varepsilon_{\ell mn}c_m^{\mu\nu}\left(\overline{\eta}_n+\frac{1}{2}\eta_n'^{\mu}\right)\right]\\
&-\varepsilon_{ijk}k\delta_{j\ell}x_k^{\mu\sigma}\left[\overline{e}_{\ell m}c_m^{\mu\sigma}-\varepsilon_{\ell mn}c_m^{\mu\sigma}\left(\overline{\eta}_n+\frac{1}{2}\eta_n'^{\mu}+\frac{1}{2}\eta_n'^{\sigma}\right)\right]\\
&+\varepsilon_{ijk}k\left(n_j^{\mu\sigma}n_\ell^{\mu\sigma}+\mu^{\mu\sigma}\overline{n}_j^{\mu\sigma}n_\ell^{\mu\sigma}\right)x_k^{\mu\sigma}\left[\overline{e}_{\ell m}c_m^{\mu\sigma}-\varepsilon_{\ell mn}c_m^{\mu\sigma}\left(\overline{\eta}_n+\frac{1}{2}\eta_n'^{\mu}+\frac{1}{2}\eta_n'^{\sigma}\right)\right]=0
\end{aligned}
\tag{8.1}
$$

The last term in this equation (which includes $\eta_n'^{\sigma}$) vanishes, because the sliding interaction does not couple to the spin. Thus, calling the angle of the sliding contact α, the solutions for the spins of the two particles is the same and equal to

$$
\begin{aligned}
\eta_3'^{\mu}=\eta_3'^{\sigma}&=-2\frac{N_c-1}{N_c-2}\overline{\eta}_3+\frac{\cos 2\alpha}{N_c-2}\left(\mu_s\left(\overline{e}_{11}-\overline{e}_{22}\right)+2\overline{e}_{12}\right)\\
&-\frac{\sin 2\alpha}{N_c-2}\left(\overline{e}_{11}-\overline{e}_{22}-2\mu_s\overline{e}_{12}\right)-\frac{\pi\left(\overline{p}_{11}-\overline{p}_{22}\right)}{N_c-2}\overline{e}_{12}+\frac{\mu_s}{N_c-2}\left(\overline{e}_{11}+\overline{e}_{22}\right)
\end{aligned}
$$

The stress is now easily evaluated

$$
\begin{aligned}
\overline{\sigma}_{ij}&=\frac{1}{2V}\sum_{\mu}\sum_{\nu=1}^{N^\mu} f_i^{\mu\nu}c_j^{\mu\nu}\\
&=\frac{1}{2V}\sum_{\mu}\sum_{\nu=1}^{N^\mu} K_{ik}^{\mu\nu}\left[\overline{e}_{k\ell}c_\ell^{\mu\nu}-\varepsilon_{k\ell m}\left(c_\ell^{\mu\nu}\overline{\eta}_m+\frac{1}{2}c_\ell^{\mu\nu}\eta_m'^{\mu}+\frac{1}{2}c_\ell^{\mu\nu}\eta_m'^{\nu}\right)\right]c_j^{\mu\nu}
\end{aligned}
$$

For the sum over the contacts of the particles with one slip the same procedure is followed that was used for the calculation of the spin

$$\overline{2v\left(\sigma_{ij}\right)_{slip} = \sum_{\nu=1}^{N^{\mu}} k\delta_{i\ell}c_j^{\mu\nu}\left[\overline{e}_{\ell m}c_m^{\mu\nu} - \varepsilon_{\ell mn}c_m^{\mu\nu}\left(\overline{\eta}_n + \frac{1}{2}\eta_n'^{\mu}\right)\right]}$$

$$\overline{+k\left(n_i^{\mu\sigma}n_\ell^{\mu\sigma} + \mu^{\mu\sigma}\overline{n}_i^{\mu\sigma}n_\ell^{\mu\sigma} - \delta_{i\ell}\right)c_j^{\mu\sigma}\left[\overline{e}_{\ell m}c_m^{\mu\sigma} - \varepsilon_{\ell mn}c_m^{\mu\sigma}\left(\overline{\eta}_n + \frac{1}{2}\eta_n'^{\mu} + \frac{1}{2}\eta_n'^{\sigma}\right)\right]}$$

The averaging bars are needed because the distribution of slipping angles needs to be specified. The result of this is more involved. First one term under the averaging bar (that is, for one particular angle of the slipping contact, α) is evaluated. The number of contacts per particle is N_c and the anisotropy index a_p is such that $(1+a_p)/(1-a_p) = \overline{p}_{11}/\overline{p}_{22}$. The $i=1, j=2$ element is equal to the $i=2, j=1$ element. The $i=1, j=1$ element in a fabric that is aligned with the coordinate axes comes out as

$$\frac{1}{2}kN_c\overline{c}^2\left(1+\frac{1}{2}a\right)\overline{e}_{11}$$
$$+\frac{1}{4}\frac{kN_c\overline{c}^2}{N_c-2}\left(\sin^2 2\alpha\left(\overline{e}_{22}-\overline{e}_{11}\right) - 2a\sin 2\alpha\left(a\overline{e}_{12}+\overline{\eta}\right) + \sin 4\alpha\overline{e}_{12}\right)$$
$$+\frac{1}{4}\frac{k\mu_s N_c\overline{c}^2}{N_c-2}\left(\sin 2\alpha\left(\overline{e}_{11}+\overline{e}_{22}\right) + \frac{1}{2}\sin 4\alpha\left(\overline{e}_{11}-\overline{e}_{22}\right) + 2\sin^2 2\alpha\overline{e}_{12}\right)$$

The $i=2, j=2$ element is similarly

$$\frac{1}{2}kN_c\overline{c}^2\left(1-\frac{1}{2}a\right)\overline{e}_{22}$$
$$+\frac{1}{4}\frac{kN_c\overline{c}^2}{N_c-2}\left(\sin^2 2\alpha\left(\overline{e}_{11}-\overline{e}_{22}\right) + 2\sin 2\alpha\left(a\overline{e}_{12}+\overline{\eta}\right) - \sin 4\alpha\overline{e}_{12}\right)$$
$$-\frac{1}{4}\frac{k\mu_s N_c\overline{c}^2}{N_c-2}\left(\sin 2\alpha\left(\overline{e}_{11}+\overline{e}_{22}\right) - \frac{1}{2}\sin 4\alpha\left(\overline{e}_{11}-\overline{e}_{22}\right) + 2\sin^2 2\alpha\overline{e}_{12}\right)$$

For the $i=1, j=2$ element the contributions that are proportional to $\overline{e}_{12}$ are given only

$$\frac{1}{8}\frac{kN_c\overline{c}^2}{N_c-2}\left(N_c\left(4-a_p^2\right)\overline{e}_{12}-12\overline{e}_{12}-2a_p\overline{\eta}\right)$$

$$+\frac{1}{2}\frac{kN_c\overline{c}^2}{N_c-2}\left(\sin^2 2\alpha\overline{e}_{12}+\frac{3}{2}a_p\cos 2\alpha\overline{e}_{12}+\frac{1}{2}\overline{\eta}\right)$$

$$+\frac{1}{4}\frac{k\mu_s N_c\overline{c}^2}{N_c-2}\left(a_p\sin 2\alpha-\sin 4\alpha\right)\overline{e}_{12}$$

The $i=2,\ j=1$ element is identical.

The result still contains the mean spin, which is determined by averaging over the slipping angles. Two special cases are considered.

8.3 Mean-field estimate with randomly distributed slip angles

The first special case pertains to the situation in which the slips are distributed randomly. In this scenario the mean value of the slip coefficient vanishes and an average may be taken over all angles. The result may be combined with the all-stick result, the fraction of particles that have one slipping contact is called ϕ_s; the fraction of particles that has no slipping contacts is $(1-\phi_s)$. The mean spin is then determined by requiring the average of the spin fluctuations to vanish. The result is

$$\overline{\eta}=-a_p\overline{e}_{12}\frac{N_c+2(\phi_s-1)}{2(N_c+\phi_s-2)}$$

Thus the mean spin depends on the shear strain increment only and is proportional to the anisotropic mean packing characteristics. The combined stresses are

$$2v(\sigma_{11})_{mean}=\frac{k\overline{c}^2N_c}{8(N_c-2)}\left[\left(2N_c\left(a_p+2\right)-4a_p-\phi_s-8\right)\overline{e}_{11}+\phi_s\overline{e}_{22}\right]$$

$$2v(\sigma_{22})_{mean}=\frac{k\overline{c}^2N_c}{8(N_c-2)}\left[\phi_s\overline{e}_{11}+\left(2N_c\left(2-a_p\right)+4a_p-\phi_s-8\right)\overline{e}_{22}\right]$$

$$2v(\sigma_{12})_{mean}=\frac{1}{4}k\overline{c}^2N_c\left(\frac{2N_c-\phi_s-4}{N_c-2}-\frac{1}{2}a_p^2\frac{N_c+2\phi_s-2}{N_c+\phi_s-2}\right)\overline{e}_{12}$$

This result is similar to the mean-field estimate for a medium with a suitably chosen tangential interaction.

8.4 Mean-field estimate with concentrated slip angles

The second case pertains to the situation in which the slips are concentrated in certain directions. This case is especially important when the sign of the friction coefficient is associated with a deviatoric stress; this case is explored in this section.

8.4.1 *A slip angle associated with the deviatoric stress*

The principal stress axes are aligned with the coordinate axes (as is the mean fabric tensor). An illustration is provided in Fig. 8.1.

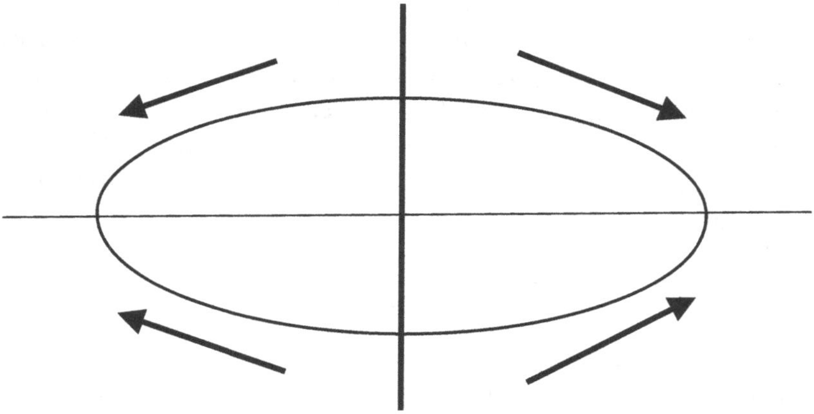

Figure 8.1. The stress ellipse and the sign of the friction coefficient $\mu_s = (F_p \bar{n}_p)/(F_k n_k) = (\Sigma_{pq} \bar{n}_p n_q)/(\Sigma_{pq} n_p n_q)$ (in this case the maximum value of the stress ratio is at an angle of approx 55^0 in the first quadrant. The arrows give an indication of the direction of the tangential force.

So, while only one sliding contact per particle is taken into account, for every angle α the conjugate angles $\alpha + \pi$, $-\alpha$ and $-\alpha + \pi$ will participate with the signs of μ_s of +, +, –, –, respectively. All these are added to the sums of the stress contributions; averaging over the angles is then done in the first quadrant according to a prescribed distribution. Here the result for a distribution of slipping angles is chosen that represents one particular angle α in the first quadrant and its three conjugate counterparts. These four have equal probability and averaging the expressions for $2v(\boldsymbol{\sigma})_{slip}$ obtained at the end of Section 8.3 with the appropriate signs for the inter-particle friction angle yields the following:

$$\overline{\eta} = \frac{2 - 4\sin^2\alpha - N_c a_p}{2(N_c - 1)}\overline{e}_{12}$$

$$2v(\sigma_{12})_{slip(\alpha)} = \frac{1}{2}N_c k\overline{c}^2\overline{e}_{12}$$
$$\times\left(\frac{\frac{1}{4}N_c\left(4 - a_p^2\right) - 2 + \sin^2 2\alpha + \cos 2\alpha}{N_c - 1} + \frac{\cos 2\alpha - \frac{1}{4}(1 + \mu_s)\sin 4\alpha}{N_c - 2}\right)$$

$$2v(\sigma_{11})_{slip(\alpha)} = \frac{1}{4}N_c k\overline{c}^2\left(a_p + 2 + \frac{\mu_s\left(\frac{1}{2}\sin 4\alpha + \sin 2\alpha\right) - \sin^2 2\alpha}{N_c - 2}\right)\overline{e}_{11}$$
$$+\frac{1}{4}N_c k\overline{c}^2\left(\frac{\mu_s\left(-\frac{1}{2}\sin 4\alpha + \sin 2\alpha\right) + \sin^2 2\alpha}{N_c - 2}\right)\overline{e}_{22}$$

$$2v(\sigma_{22})_{slip(\alpha)} = \frac{1}{4}N_c k\overline{c}^2\left(\frac{\mu_s\left(-\frac{1}{2}\sin 4\alpha - \sin 2\alpha\right) + \sin^2 2\alpha}{N_c - 2}\right)\overline{e}_{11}$$
$$+\frac{1}{4}N_c k\overline{c}^2\left(2 - a_p + \frac{\mu_s\left(\frac{1}{2}\sin 4\alpha - \sin 2\alpha\right) - \sin^2 2\alpha}{N_c - 2}\right)\overline{e}_{22}$$

These formulas are not directly very transparent, which is why a graphical illustration is supplied in Fig. 8.2.

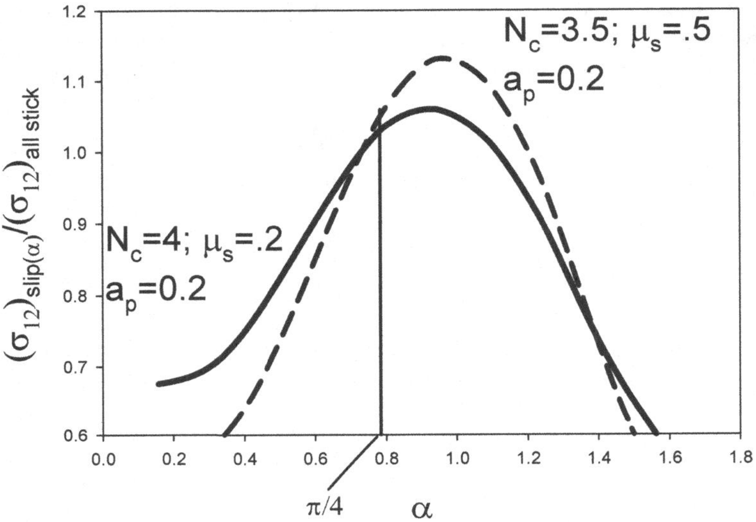

Figure 8.2. The ratio $(\sigma_{12})_{slip(\alpha)}/(\sigma_{12})_{all\,stick}$ for an assembly of particles with one slipping contact each in the mean field approximation. Parameters as indicated in the graph.

The shear modulus is shown, normalised to the all-stick case. The parameters — friction coefficient, anisotropy index and number of contacts — have very little impact on the result; the shear modulus is relatively unaffected by the slip phenomenon when the slip angle lies in the vicinity of $\pi/4$. Outside this region there is a reduction in the shear modulus.

The normal moduli are most easily studied in terms of the ratio $\delta = E_{2211}/E_{2222}$ and the non-dimensional determinant $\Delta = (2\nu)^2 (E_{1111}E_{2222} - E_{1122}E_{2211})/(k^2\overline{c}^4)$. The former gives an indication of the dilatancy sensitivity (when the ratio exceeds unity the assembly expands in a biaxial cell test); the latter points to the stability of the assembly (when Δ approaches zero the assembly exhibits a possible rupture layer).

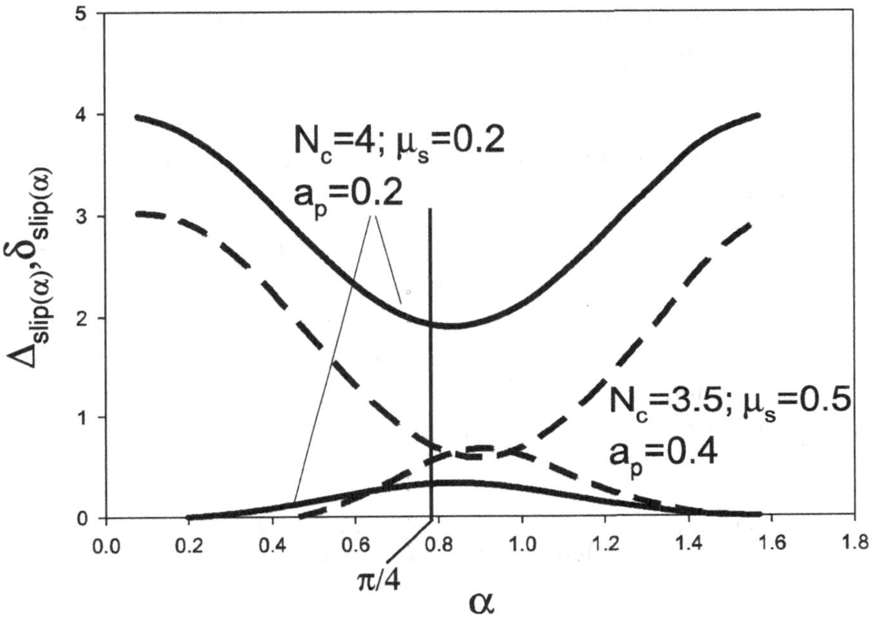

Figure 8.3. The stiffness determinant $\Delta_{slip(\alpha)}$ (top) and the dilatancy ratio $\delta_{slip(\alpha)}$ (bottom) as a function of the slip angle α. The parameters are indicated in the graph; the solid lines are for $N_c = 4; a_p = 0.2; \mu_s = 0.2$; the dashed lines for $N_c = 3.5; a_p = 0.4; \mu_s = 0.5$.

Interesting features come to the fore. The isostatic limit for one slipping contact for all particles in two dimensions is at $N_c = 3.5$. Well-away from this value and moderate values for the anisotropy and the friction coefficient, the assembly is neither dilatant nor prone to rupture layer formation. Decreasing the number of contacts and increasing both the anisotropy and the friction coefficient leads to a more dilatant assembly and also one that is closer to rupture layer formation. All this takes place at a certain angle for the slipping contact. This trend is what may be expected when studying the data from numerical simulations, although it is emphasised that the mean-field approximation is a very rough approximation and the assumption that all particles have equal numbers of contacts and that only one of them slips is definitely artificial. Nevertheless, the key physical features show themselves.

8.4.2 *Further investigation of the slip angle and fabric heterogeneity*

The question is now whether there is any favoured angle for a direction of a slipping contact. Insight may come from the application of Equation (5.1), which may be used to calculate — in an average way — the value of the force ratio in any direction. Simple algebra permits the expression of the contact force ratio μ_c as a function of the contact angle α for given fabric anisotropy a_p and stress ratio $R = \Sigma_{11} / \Sigma_{22}$.

$$\mu_c(\alpha) = \frac{R(a_p - 2) + a_p + 2}{R(a_p - 2)\cos^2\alpha - (a_p + 2)\sin^2\alpha} \sin\alpha\cos\alpha \tag{8.2}$$

This relationship is plotted below in Fig. 8.4 for two values of a_p and $R = 4$. Now, the value of the force ratio cannot exceed the material constant μ_s and therefore the slip angle is just in the maximum of the graph. It is observed that a higher value of the maximum slip angle can be achieved at the same stress ratio for smaller values of the fabric anisotropy.

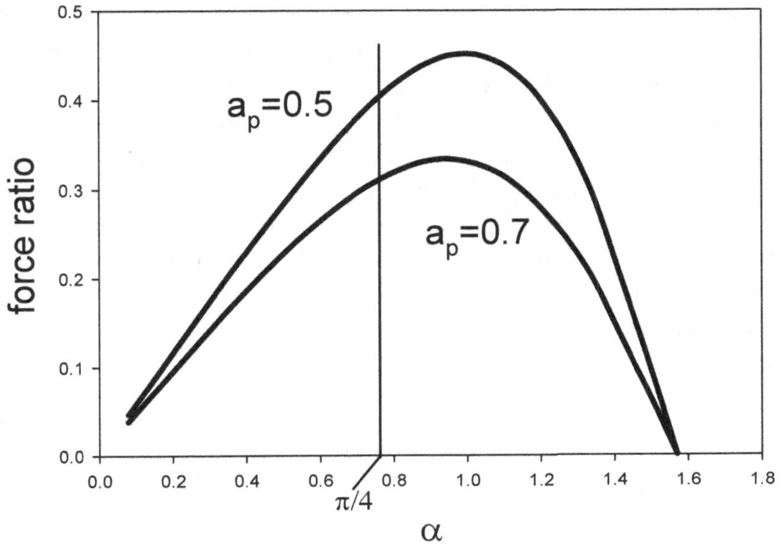

Figure 8.4. The force ratio as a function of the angle of the contact for $R = 4$ and two values for the anisotropy parameter a_p.

Conversely, for a given material parameter, the inter-particle friction ratio μ_s, the anisotropy required can be expressed in the stress ratio. To find that relationship the maximum of Equation (8.1) is found. This is easily obtained; the critical angle is

$$\alpha_s = \cot^{-1}\left(\frac{\sqrt{a_p + 2}}{\sqrt{R(2 - a_p)}}\right)$$

The associated anisotropy parameter is

$$a_p\Big|_{\alpha=\alpha_s} = 2\frac{R^2 - 4R\mu_s\sqrt{\mu_s^2 + 1} - 1}{R^2 + 2R(2\mu_s^2 + 1) + 1}$$

The narrative that describes the progress of the assembly as the stress ratio is gradually increased is now becoming clear. As deviatoric strain is applied the assembly becomes more anisotropic, while slipping contacts develop. These will be increasingly concentrated in the direction of α_s. Then — consulting Fig. 8.3 — the dilatancy ratio will increase (and could even reach values greater than unity, implying expansion of the assembly in a biaxial cell test) and the value of the scaled determinant Δ will drop to zero, which results in the overall stress ratio remaining constant.

These features are observed in both physical and numerical experiments. The remarkable thing is that they follow from a relatively simple mean-strain theory. The theory is augmented by making allowance for *fabric heterogeneity*. Thus, as far as the effects of slipping contacts is concerned a mean-field theory is still used, but a first-order estimate of the influence of fabric heterogeneity is readily introduced by incorporating the formulas of Section 7.7 as an extra perturbation. Essentially, the principal moduli are affected only, with the result

$$2v(\sigma_{11})_{slip(\alpha)} = \frac{1}{4}N_c k\bar{c}^2\left[a_p + 2 + 2N_V\frac{(P')^2}{\bar{p}^2}\frac{\bar{c}}{\hat{a}}S_1\left(\frac{\bar{c}}{\hat{a}}\right)\left(\frac{\sqrt{2-a_p}}{a_p\sqrt{2+a_p}} - \frac{1}{a_p}\right)\right]\bar{e}_{11}$$
$$+\frac{1}{4}N_c k\bar{c}^2\left(\frac{\mu_s\left(\frac{1}{2}\sin 4\alpha + \sin 2\alpha\right) - \sin^2 2\alpha}{N_c - 2}\right)\bar{e}_{11}$$
$$+\frac{1}{4}N_c k\bar{c}^2\left(+\frac{\mu_s\left(-\frac{1}{2}\sin 4\alpha + \sin 2\alpha\right) + \sin^2 2\alpha}{N_c - 2}\right)\bar{e}_{22} \tag{8.3}$$

$$2v(\sigma_{22})_{slip(\alpha)} = -\frac{1}{4}N_c k\bar{c}^2\frac{\mu_s\left(\frac{1}{2}\sin 4\alpha + \sin 2\alpha\right) - \sin^2 2\alpha}{N_c - 2}\bar{e}_{11}$$
$$+\frac{1}{4}N_c k\bar{c}^2\left(2 - a_p + 2N_V\frac{(P')^2}{\bar{p}^2}\frac{\bar{c}}{\hat{a}}S_1\left(\frac{\bar{c}}{\hat{a}}\right)\left(\frac{1}{a_p} - \frac{\sqrt{2+a_p}}{a_p\sqrt{2-a_p}}\right)\right)\bar{e}_{22}$$
$$+\frac{1}{4}N_c k\bar{c}^2\left(\frac{\mu_s\left(\frac{1}{2}\sin 4\alpha - \sin 2\alpha\right) - \sin^2 2\alpha}{N_c - 2}\right)\bar{e}_{22} \tag{8.4}$$

It is now possible to find the point at which the outer determinant vanishes (which corresponds to peak stress ratio) and as a function of the number of contacts and the heterogeneity parameter $(P')^2/\overline{p}^2$ the associated anisotropy parameter a_p, dilatancy ratio δ and the stress ratio $R = \Sigma_{11}/\Sigma_{22}$ follow. All these can be read at the point for which $\alpha = \alpha_s$. It is then assumed that the vast majority of sliding contacts are in the vicinity of this angle (and its three conjugate counterparts). The results are plotted in Fig. 8.5.

It is necessary to add a note on the heterogeneity parameter. Firstly, in the above equations it has been assumed that *all* particles have *one* sliding contact. That is a simplification, because there may well be particles with no slipping contacts at all. These would affect the result by increasing the principal moduli. Also, there may be a fraction of particles with no contacts at all and while this is accounted for in the heterogeneity parameter, it is difficult to ascertain how exactly these particles affect the analysis based on an explicit sliding contact. Secondly, it was seen that the heterogeneity acts in concert with the influence function $(\overline{c}/\widehat{a})S_1$. These two may as well be taken together into one heterogeneity parameter:

$$\widehat{H} = \frac{(P')^2}{\overline{p}^2}\frac{\overline{c}}{\widehat{a}}S_1\left(\frac{\overline{c}}{\widehat{a}}\right)$$

In fact, this parameter acts as an amalgam, incorporating many aspects, such as particles having no contacts, particles with no slipping contacts, as well as fabric variations. Therefore, this parameter is a collective correction parameter, as it is not convenient to add extra variables to the theory. It must be borne in mind that this is a rather primitive mean-field theory that essentially has the purpose of elucidating the physics of the problem.

Despite the simplicity of the theory some main features come to the fore. When there are few contacts — that is for a loosely packed material — the stress ratio at peak, the anisotropy parameter and the dilatancy ratio are smaller than when there are more contacts. When the material is loosely packed it is plausible that the heterogeneity parameter is also larger; this regime is represented by the solid lines in the graph and it is observed that this leads to smaller stress ratios, less anisotropy and smaller dilatancies. All these effects, qualitatively at least, have also been observed in both physical and numerical experiments. It is noted that these realistic results are obtained by taking into account a small amount of fabric heterogeneity. A further result is that the dilatancy exhibits a plateau for small contact numbers.

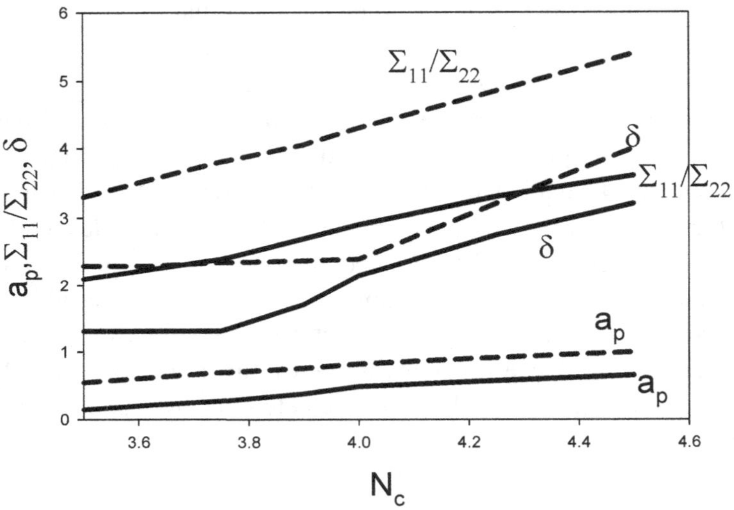

Figure 8.5. The anisotropy parameter, the stress ratio and the dilatancy parameter as functions of the mean number of contacts per particle. The solid lines correspond to a heterogeneity parameter of $\widehat{H} = 0.1$ and the dashed lines to $\widehat{H} = 0.05$. All calculations done at peak stress ratio when $\Delta = 0$.

8.4.3 *Verification of the friction criterion*

One aspect that has so far been tacitly circumvented is the peculiar requirement of the frictional interaction that a verification is carried out to check that the displacement path of the sliding contact, as applied to the sticking rather than a sliding state of the interaction, does not lead to a lower stress ratio (which would make it a sticking contact once again). The increase in stress ratio is easily obtained, it is

$$\frac{1}{F_\perp}\left(f_\parallel - \mu_s f_\perp\right)$$

The compressive force $F_\perp$ is always positive. A diagonal sticking interaction has been used and therefore the increase in terms of displacement becomes

$$\frac{k}{F_\perp}\left(d_\parallel - \mu_s d_\perp\right)$$

The outcome obviously depends on the mean-strain increment that is chosen. Here, an illustration is provided for a zero shear strain. The mean spin is then also zero and the increment of increase in stress ratio is

$$\frac{k}{2F_\perp}\frac{N_c \mu_s}{N_c - 2}\left(\overline{e}_{22} - \overline{e}_{11}\right)\cos 2\alpha_s$$

$$-\frac{k}{2F_\perp}\left[\frac{N_c}{N_c - 2}\left(\overline{e}_{22} - \overline{e}_{11}\right)\sin 2\alpha_s + \frac{\mu_s}{N_c - 2}\left(\overline{e}_{22} + \overline{e}_{11}\right)\right]$$

It is easy to verify that for the choice of parameters for the inter-particle friction, the number of contacts and the slip angle broadly in the range given above, this is always strongly positive while $\overline{e}_{11} > 0, \overline{e}_{22} < 0$. This is largely due to the extra particle spin that is incurred, which is enabled by the sliding. A similar calculation shows that for contacts that are in the quadrants that are adjacent to the one in which the slip takes place the incremental increase in the force ratio is very small and may even be negative. For contacts in the opposite quadrant the increase in the tangential displacement is substantially smaller than the one for the sliding contact.

In this calculation the fabric heterogeneity has been overlooked. The result for the mean-field calculation is emphatic however: sliding contacts for a dilatant strain path persist. This is not to say that the mean-field calculation is particularly accurate; in the next section refinements will be introduced.

8.5 Strain fluctuations

An obvious improvement on the mean strain theory is to consider strain fluctuations. The purpose of this analysis is to investigate the effects of friction and therefore fabric fluctuations are ignored for the moment. These will be added later as an extra perturbation.

The calculation is somewhat involved. There are a number of stages. The first is the determination of the spins while allowing for strain fluctuations. The second stage is the calculation of the displacement fluctuations, taking into account the fact that for sliding interactions a *particle pair* must be considered, rather than in single particle. The third stage involves an estimate of the strain fluctuations, derived from the displacement fluctuations, necessary to evaluate the spin fluctuations. Then all the elements are in place to work out the assembly-averaged incremental stiffness components. Finally, fabric heterogeneity is again added as a refinement. Other possible refinements are also discussed.

8.5.1 *Determining the spins*

Starting point is the determination of the spins. In the moment equation (8.1) the mean strain is replaced by the mean strain plus fluctuations. So, instead of the mean strain, the local strain is inserted

$$
\begin{aligned}
&\varepsilon_{ijk}\sum_{\nu=1}^{N^\mu} k\delta_{j\ell}x_k^{\mu\nu}\left[e_{\ell m}^{\mu}c_m^{\mu\nu}-\varepsilon_{\ell mn}c_m^{\mu\nu}\left(\overline{\eta}_n+\frac{1}{2}\eta_n'^{\mu}\right)\right]\\
&-\varepsilon_{ijk}k\delta_{j\ell}x_k^{\mu\sigma}\left[e_{\ell m}^{\mu}c_m^{\mu\sigma}-\varepsilon_{\ell mn}c_m^{\mu\sigma}\left(\overline{\eta}_n+\frac{1}{2}\eta_n'^{\mu}+\frac{1}{2}\eta_n'^{\sigma}\right)\right]\\
&+\varepsilon_{ijk}k\left(n_j^{\mu\sigma}n_\ell^{\mu\sigma}+\mu^{\mu\sigma}\overline{n}_j^{\mu\sigma}n_\ell^{\mu\sigma}\right)x_k^{\mu\sigma}\left[e_{\ell m}^{\mu}c_m^{\mu\sigma}-\varepsilon_{\ell mn}c_m^{\mu\sigma}\left(\overline{\eta}_n+\frac{1}{2}\eta_n'^{\mu}+\frac{1}{2}\eta_n'^{\sigma}\right)\right]=0
\end{aligned}
$$

A concomitant equation for particle σ has a similar form. The solution is entirely analogous to the mean-field case

$$\eta_3'^{\mu} = \eta_3'^{\sigma} = -2\frac{N_c - 1}{N_c - 2}\bar{\eta}_3 + \frac{\cos 2\alpha}{N_c - 2}\left(\mu_s\left(e_{11}^{\mu} - e_{22}^{\mu}\right) + 2e_{12}^{\mu}\right)$$
$$-\frac{\sin 2\alpha}{N_c - 2}\left(e_{11}^{\mu} - e_{22}^{\mu} - 2\mu_s e_{12}^{\mu}\right) - \frac{\pi\left(\bar{p}_{11} - \bar{p}_{22}\right)}{N_c - 2}e_{12}^{\mu} + \frac{\mu_s}{N_c - 2}\left(e_{11}^{\mu} + e_{22}^{\mu}\right)$$

The average of this expression must vanish. Inspection of averages and fluctuations under the two scenarios discussed in Sections 8.3 and 8.4 shows that the terms proportional to $\mu_s \cos 2\alpha$, $\sin 2\alpha$ and μ_s all have zero mean. As a result the spin fluctuations are

$$\eta_3'^{\mu} = \eta_3'^{\sigma} = \left(\frac{\cos 2\alpha}{N_c - 2}\mu_s - \frac{\sin 2\alpha}{N_c - 2}\right)\left(\bar{e}_{11} - \bar{e}_{22}\right) + \frac{\mu_s}{N_c - 2}\left(\bar{e}_{11} + \bar{e}_{22}\right)$$
$$-\frac{\pi\left(\bar{p}_{11} - \bar{p}_{22}\right)}{N_c - 2}e_{12}'^{\mu} + 2\overline{\left(\frac{\cos 2\alpha}{N_c - 2} + 2\mu_s\frac{\sin 2\alpha}{N_c - 2}\right)}e_{12}'^{\mu}$$

Therefore, it becomes necessary to obtain an estimate for the shear strain fluctuations and this aspect is also the one that distinguishes it from the mean strain approximation.

8.5.2 *The displacement fluctuations associated with a sliding neighbouring particle derived from the force equilibrium equations*

The force equilibrium equations are

$$\sum_{\nu=1}^{N^{\mu}} K_{ij}^{\mu\nu}\left[e_{j\ell}^{\mu}c_{\ell}^{\mu\nu} + \frac{1}{2}\frac{\partial^2 u_j}{\partial x_k \partial x_\ell}\bigg|^{\mu} c_k^{\mu\nu}c_{\ell}^{\mu\nu} - \varepsilon_{jk\ell}\left(c_k^{\mu\nu}\bar{\eta}_\ell + \frac{1}{2}c_k^{\mu\nu}\eta_\ell'^{\mu} + \frac{1}{2}c_k^{\mu\nu}\eta_\ell'^{\nu}\right)\right] = 0$$

Averages and fluctuations may be distinguished. The double displacement gradient is a fluctuating quantity and all the odd structural sums ditto, so the first order fluctuation terms together are

$$\sum_{\nu=1}^{N^{\mu}} K_{ij}^{\mu\nu}\left(\bar{e}_{j\ell}c_{\ell}^{\mu\nu} - \varepsilon_{jk\ell}c_k^{\mu\nu}\bar{\eta}_\ell\right) + \frac{1}{2}\frac{\partial^2 u_j}{\partial x_k \partial x_\ell}\bigg|^{\mu}\overline{\sum_{\nu=1}^{N^{\mu}} K_{ij}^{\mu\nu}c_k^{\mu\nu}c_{\ell}^{\mu\nu}} = 0$$

The fluctuating structural sum can be written as a derivative, that is

$$\sum_{\nu=1}^{N^\mu} K_{ij}^{\mu\nu} c_\ell^{\mu\nu} \approx \frac{1}{2} N_V \frac{\partial}{\partial x_k} \left(\sum_{\nu=1}^{N^\mu} K_{ij}^{\mu\nu} c_\ell^{\mu\nu} c_k^{\mu\nu} - \overline{\sum_{\nu=1}^{N^\mu} K_{ij}^{\mu\nu} c_\ell^{\mu\nu} c_k^{\mu\nu}} \right)$$

The fluctuational content in this case derives from the slipping contact, which leads to the form

$$A'^{\mu}_{ijk\ell} = \sum_{\nu=1}^{N^\mu} K_{ij}^{\mu\nu} c_\ell^{\mu\nu} c_k^{\mu\nu} - \overline{\sum_{\nu=1}^{N^\mu} K_{ij}^{\mu\nu} c_\ell^{\mu\nu} c_k^{\mu\nu}} \approx k\left(n_i^{\mu\sigma} n_j^{\mu\sigma} + \mu_s^{\mu\sigma} \overline{n}_i^{\mu\sigma} n_j^{\mu\sigma} - \delta_{ij}\right) c_k^{\mu\sigma} c_\ell^{\mu\sigma}$$
$$- \overline{k\left(n_i^{\mu\sigma} n_j^{\mu\sigma} + \mu_s^{\mu\sigma} \overline{n}_i^{\mu\sigma} n_j^{\mu\sigma} - \delta_{ij}\right) c_k^{\mu\sigma} c_\ell^{\mu\sigma}}$$

The whole set of equations that was used in the previous chapters can be re-deployed. However, there is now a detail that needs to be addressed. As was observed in the previous section, the fluctuation that arises from particle μ cannot be viewed independently from the one that is associated with particle σ, with which it shares a sliding contact. Starting point is once more Equation (6.3). Noting the modifications made in the previous chapter the term $\mathbf{A}'(\overline{\mathbf{e}} - \boldsymbol{\varepsilon}\overline{\boldsymbol{\eta}})$ is again replaced by $\mathbf{Z}'$. The displacement fluctuation takes the form

$$u'_a(\mathbf{x}) = \frac{i}{(2\pi)^2} \int d_2 k e^{i\mathbf{k}.\mathbf{x}} P_{ai}^{-1} k_j \int d_2 y Z'_{ij}(\mathbf{y}) e^{-i\mathbf{k}.\mathbf{y}} \tag{8.5}$$

The fluctuations $\mathbf{Z}'$ now have two contributions; the first – originating from particle μ and centred on the origin – is treated along the same lines as before, but the second one stems from the fluctuating field associated with the neighbour σ with which it shares a frictional sliding contact. The analysis needs to be extended to accommodate the latter. The result for the displacement difference between the two particles that share a sliding contact is given in Equation (8.6).

The analysis goes as follows.

The centre of particle σ is located at position $\mathbf{c}^{\mu\sigma}$ and the fluctuating field $\mathbf{Z}'$ will be assumed to be 'smeared out' in radial fashion as $Z'_{ij}(\mathbf{c}^{\mu\sigma})\exp\left(-\left|\mathbf{c}^{\mu\sigma} - \mathbf{y}\right|^2 / \tilde{a}^2\right)$.

The following integral must now be evaluated

$$\int d_2 y Z'_{ij}(\mathbf{y}) e^{-i\mathbf{k}.\mathbf{y}} = Z'_{ij}(\mathbf{c}^{\mu\sigma}) \int d_2 y \exp\left(-\left|\mathbf{c}^{\mu\sigma} - \mathbf{y}\right|^2 / \widehat{a}^2\right) e^{-i\mathbf{k}.\mathbf{y}}$$

A coordinate transform $\mathbf{z} = \mathbf{y} - \mathbf{c}^{\mu\sigma}$ leads to

$$\int d_2 y Z'_{ij}(\mathbf{y}) e^{-i\mathbf{k}.\mathbf{y}} = Z'_{ij}(\mathbf{c}^{\mu\sigma}) e^{-i\mathbf{k}.\mathbf{c}^{\mu\sigma}} \int d_2 z \exp\left(-z^2 / \widehat{a}^2\right) e^{-i\mathbf{k}.\mathbf{z}}$$

This integral is again quite easily done and the result is

$$\pi \widehat{a}^2 \exp\left(-\frac{1}{4}\widehat{a}^2 k^2\right) Z'_{ij}(\mathbf{c}^{\mu\sigma}) e^{-i\mathbf{k}.\mathbf{c}^{\mu\sigma}}$$

So, the contribution of this fabric fluctuation to the fluctuating displacement field (8.5) turns out to be

$$\begin{aligned} u'_a(\mathbf{x}) &= \frac{i\widehat{a}^2}{4\pi} \int_0^{2\pi} d\psi e^{i\mathbf{k}.\mathbf{x}} \int_0^{\infty} dk k P^{-1}_{ai} k_j \exp\left(-\frac{1}{4}\widehat{a}^2 k^2\right) Z'_{ij}(\mathbf{c}^{\mu\sigma}) e^{-i\mathbf{k}.\mathbf{c}^{\mu\sigma}} \\ &= \frac{i\widehat{a}^2}{4\pi} \int_0^{2\pi} d\psi e^{i\mathbf{k}.(\mathbf{x}-\mathbf{c}^{\mu\sigma})} \int_0^{\infty} dk k P^{-1}_{ai} k_j \exp\left(-\frac{1}{4}\widehat{a}^2 k^2\right) Z'_{ij}(\mathbf{c}^{\mu\sigma}) \end{aligned}$$

The result of the analysis of the fluctuation at the origin (that is the fluctuation centered at particle μ) may now be taken over — Section (6.5) — with the substitution $x \to \left|\mathbf{x} - \mathbf{c}^{\mu\sigma}\right|$ and the vector $\mathbf{m}$ the unit vector that points from the centre of particle σ. To distinguish this vector from the unit vector that points away from the origin at particle μ it will be called $\mathbf{m}^{(\sigma)}$; the vector $\mathbf{m}$ itself, the one pointing away from the centre of particle μ is — for clarity — denoted as $\mathbf{m}^{(\mu)}$.

For the form of $\mathbf{A}'^{(\sigma)}$ the angle of the unit normal is increased by π, thereby incurring a minus sign. As the unit normals always appear in pairs the form is the same as for $\mathbf{A}'^{(\mu)}$.

The form for the displacement fluctuation is recalled

$$\begin{aligned} u'_a(\mathbf{x}) = &-\frac{\widehat{a}}{8} p_1 \left[S_1\left(\frac{x}{\widehat{a}}\right) + S_3\left(\frac{x}{\widehat{a}}\right)\right]\left(m_i \delta_{aj} + m_j \delta_{ai} + m_a \delta_{ij}\right) Z'_{ij}(0) \\ &+ \frac{\widehat{a}}{2} p_1 \left(S_3\left(\frac{x}{\widehat{a}}\right) m_i m_j m_a\right) Z'_{ij}(0) - \frac{\widehat{a}}{2} p_0 m_j \delta_{ai} S_1\left(\frac{x}{\widehat{a}}\right) Z'_{ij}(0) \end{aligned}$$

Now, for the case of the normal and tangential interaction being equal $p_1 = 0$ and $p_0 = 1/\left(\pi \bar{c}^2 \bar{p} k\right)$; this is not entirely the case here, as there are sliding contacts. However, the p_0 term is dominant and the p_1 term is rather smaller, so approximately it follows that

$$u_a'(\mathbf{x}) \simeq -\frac{\hat{a}}{2\pi \bar{c}^2 k \bar{p}} \left[m_j^{(\mu)} S_1\left(\frac{x}{\hat{a}}\right) + m_j^{(\sigma)} S_1\left(\frac{\left|\mathbf{x} - \mathbf{c}^{\mu\sigma}\right|}{\hat{a}}\right) \right] Z_{aj}'(0)$$

Importantly, the displacement difference between the two particles μ and σ is

$$u_a'^{\sigma} - u_a'^{\mu} \simeq -\frac{\hat{a} m_j^{\mu\sigma}}{\pi \bar{c}^2 k \bar{p}} S_1\left(\frac{\bar{c}}{\hat{a}}\right) Z_{aj}'(0) \tag{8.6}$$

The way is now clear to assess the overall stress increment.

8.5.3 *The stress contribution associated with the fluctuations due to sliding; strain fluctuations and spin fluctuations*

In order to arrive at the assembly-averaged stress increment, the following expression needs to be evaluated

$$\begin{aligned}
\bar{\sigma}_{st} = &\frac{1}{2v} \overline{\sum_{\nu=1}^{N^\bullet} k \delta_{su} c_t^{\bullet\nu} \left[\bar{e}_{uv} c_v^{\bullet\nu} + u_u'^{\nu} - u_u'^{\bullet} - \varepsilon_{uk\ell} c_k^{\bullet\nu} \left(\bar{\eta}_\ell + \tfrac{1}{2}\eta_\ell'^{\bullet} + \tfrac{1}{2}\eta_\ell'^{\nu} \right) \right]} \\
&+ \frac{1}{2v} \overline{k \left(n_s^{\bullet\sigma} n_u^{\bullet\sigma} + \mu_s \bar{n}_s^{\bullet\sigma} n_u^{\bullet\sigma} - \delta_{su} \right) c_t^{\bullet\sigma} \left(\bar{e}_{uv} c_v^{\bullet\sigma} + u_u'^{\sigma} - u_u'^{\bullet} \right)} \\
&- \frac{1}{2v} \varepsilon_{uk\ell} \overline{k \left(n_s^{\bullet\sigma} n_u^{\bullet\sigma} + \mu_s \bar{n}_s^{\bullet\sigma} n_u^{\bullet\sigma} - \delta_{su} \right) c_t^{\bullet\sigma} c_k^{\bullet\sigma} \left(\bar{\eta}_\ell + \tfrac{1}{2}\eta_\ell'^{\bullet} + \tfrac{1}{2}\eta_\ell'^{\sigma} \right)}
\end{aligned}$$

In the first term the sum is again replaced by an integral and the sum over the fluctuating spins of the neighbours of $\bullet$ are neglected. Note then that the displacement fluctuations are all proportional to $\mathbf{Z}'$, which average to zero. The spin fluctuations will be required to average to zero, so these need not be calculated. Thus, the first term becomes just like the mean-field approximation

$$\frac{1}{2v}\overline{\sum_{\nu=1}^{N^{\bullet}} k\delta_{su} c_t^{\bullet\nu}\left(\overline{e}_{uk} c_k^{\bullet\nu} - \varepsilon_{uk\ell} c_k^{\bullet\nu}\overline{\eta}_\ell\right)}$$

The third term can also be simplified. As the vectors $\mathbf{c}$ and $\mathbf{n}$ both point in the same direction, it follows from the anti-symmetric properties of the Levi-Civita tensor that the spin terms of the slipping contacts vanish. The second and third terms term are therefore

$$\frac{1}{2v}\overline{k\left(n_s^{\bullet\sigma} n_u^{\bullet\sigma} + \mu_s \overline{n}_s^{\bullet\sigma} n_u^{\bullet\sigma} - \delta_{su}\right) c_t^{\bullet\sigma}\left(\overline{e}_{uv} c_v^{\bullet\sigma} + u_u'^{\sigma} - u_u'^{\bullet}\right)}$$
$$+\frac{1}{2v}\overline{k\varepsilon_{sk\ell} c_k^{\bullet\sigma} c_t^{\bullet\sigma}\left(\overline{\eta}_\ell + \frac{1}{2}\eta_\ell'^{\bullet} + \frac{1}{2}\eta_\ell'^{\sigma}\right)}$$

An estimate is needed for the strain fluctuations, which is less simple in this case because of the contribution of the neighbouring particle that slips. The displacement fluctuation is expressed in polar coordinates

$$u_a'(\mathbf{x}) \simeq -\frac{\hat{a}}{2\pi\overline{c}^2 k\overline{p}}$$
$$\times\left[m_j^{(\mu)} S_1\left(\frac{r\left|\mathbf{m}^{(\mu)}\right|}{\hat{a}}\right) + \frac{rm_j^{(\mu)} - c_j^{\mu\sigma}}{\left|r\mathbf{m}^{(\mu)} - \mathbf{c}^{\mu\sigma}\right|} S_1\left(\frac{\left|r\mathbf{m}^{(\mu)} - \mathbf{c}^{\mu\sigma}\right|}{\hat{a}}\right)\right] Z_{aj}'(0)$$

The first derivative in $\mathbf{x} = 0$ of the second term in the square brackets takes the form

$$\mathbf{V} \rightarrow \begin{pmatrix} \begin{array}{l} \frac{1}{2}S_1'\left(\frac{c}{\hat{a}}\right) + \frac{\hat{a}}{2c}S_1\left(\frac{c}{\hat{a}}\right) + \\ \frac{1}{2}\left[-\frac{\hat{a}}{c}S_1\left(\frac{c}{\hat{a}}\right) + S_1'\left(\frac{c}{\hat{a}}\right)\right]\cos 2\alpha \end{array} & \frac{1}{2}\left(S_1'\left(\frac{c}{\hat{a}}\right) - \frac{\hat{a}}{c}S_1\left(\frac{c}{\hat{a}}\right)\right)\sin 2\alpha \\ \frac{1}{2}\left(S_1'\left(\frac{c}{\hat{a}}\right) - \frac{\hat{a}}{c}S_1\left(\frac{c}{\hat{a}}\right)\right)\sin 2\alpha & \begin{array}{l} S'(0) + \frac{1}{2}S_1'\left(\frac{c}{\hat{a}}\right) + \frac{\hat{a}}{2c}S_1\left(\frac{c}{\hat{a}}\right) + \\ \frac{1}{2}\left[\frac{\hat{a}}{c}S_1\left(\frac{c}{\hat{a}}\right) - S_1'\left(\frac{c}{\hat{a}}\right)\right]\cos 2\alpha \end{array} \end{pmatrix}$$

An estimate of the strain fluctuation is obtained by means of a least squares procedure. While in principle it is possible to obtain an estimate of the strain fluctuation by simply taking the derivative of the displacement fluctuation in $\mathbf{x}=0$, this does not lead to a satisfactory procedure for two reasons. Firstly, the strain fluctuation should contain information about the neighbouring particles and these are all located at more or less the same distance from the centre of particle μ; therefore, a variation in the distance (as is implied by taking the gradient in $\mathbf{x}=0$) does not yield any useful information derived from the first term in square brackets. Obviously, the second term does imply a variation in the distance, because the origin of this term is not located in the centre of particle μ. Secondly, the presence of the second term in the square brackets contains information that is specific to the angle α. This information correlates with information in $\mathbf{Z}'$ and, whereas the average of $\mathbf{Z}'$ is zero, the average of the product of the contribution to the strain fluctuation does not necessarily vanish. This would make it *not* a fluctuating quantity.

In order to remedy these two objections a least squares procedure is advocated, which — because of its functional form — enables the imposition of a stipulation that forces the mean of the gradient to vanish. The functional that delivers this goal is easily constructed. It takes the form

$$\sum_{\mu}\sum_{\nu=1}^{N^\mu}\left[e_{ij}'^{\mu}c_j^{\mu\nu}-u_i'^{\mu}\left(\mathbf{c}^{\mu\nu}\right)+u_i'^{\mu}\left(0\right)\right]^2+\lambda_{ij}\sum_{\mu}e_{ij}'^{\mu},$$

which is minimised under the condition that $\dfrac{1}{N}\sum_{\mu}e_{ij}'^{\mu}=0$ and the set $\boldsymbol{\lambda}$ are the appropriate Lagrange multipliers.

The result of the minimisation is

$$e_{i\ell}'^{\mu}=\left[\left(\sum_{\nu=1}^{N^\mu}\mathbf{cc}\right)^{-1}\right]_{\ell j}^{\mu}\sum_{\nu=1}^{N^\mu}\left[u_i'^{\mu}\left(\mathbf{c}^{\mu\nu}\right)-u_i'^{\mu}\left(0\right)\right]c_j^{\mu\nu}-\frac{1}{2}\lambda_{ij}\left[\left(\sum_{\nu=1}^{N^\mu}\mathbf{cc}\right)^{-1}\right]_{\ell j}^{\mu}$$

Replacing the sum over the branch vectors by its average and evaluating this as an integral gives

$$\left[\left(\sum_{\nu=1}^{N^\mu}\mathbf{cc}\right)^{-1}\right]_{\ell j}^{\mu} \simeq \frac{2}{N_V\overline{c}^2}\delta_{\ell j}$$

In a similar way the sum over the displacement fluctuations may be obtained. There are two contributions; the first one is simply

$$-\frac{\widehat{a}}{2\pi\overline{c}^2 k\overline{p}}S_1\left(\frac{\overline{c}}{\widehat{a}}\right)Z'_{ik}(0)\left(\frac{2}{N_V\overline{c}^2}\delta_{\ell j}\right)\sum_{\nu=1}^{N^\mu}c_j^{\mu\nu}m_k^{\mu\nu} \rightarrow -\frac{\widehat{a}}{2\pi\overline{c}^3 k\overline{p}}S_1\left(\frac{\overline{c}}{\widehat{a}}\right)Z'_{ij}(0),$$

where use was made of the approximation $\sum_{\nu=1}^{N^\mu}\mathbf{c}^{\mu\nu} \approx 0$.

For the second contribution — that is the one due to the source at particle σ — the function is approximated by expanding in a Taylor series up to first order around $\mathbf{x}=0$; this yields a similar term in the unit vector of the branch as in the first term, but its front factor is the matrix $\mathbf{V}$, as determined above. Altogether the strain fluctuation takes the form

$$\frac{\partial u'_i}{\partial x_\ell} = -\frac{Z'_{ij}(0)}{2\pi\overline{c}^2 k\overline{p}}\left(V_{j\ell} + \frac{\widehat{a}}{\overline{c}}S_1\left(\frac{\overline{c}}{\widehat{a}}\right)\delta_{j\ell}\right) + \lambda_{i\ell}$$

The factors that are needed to evaluate this are plotted in Fig. 8.6.

The Lagrange multipliers are such that the average vanishes and therefore the outcome is

$$\frac{\partial u'_i}{\partial x_\ell} = -\frac{Z'_{ij}(0)}{2\pi\overline{c}^2 k\overline{p}}\left(V_{j\ell} + \frac{\widehat{a}}{\overline{c}}S_1\left(\frac{\overline{c}}{\widehat{a}}\right)\delta_{j\ell}\right) + \overline{\frac{Z'_{ij}(0)V_{j\ell}}{2\pi\overline{c}^2 k\overline{p}}}$$

It is noted that the terms of $\mathbf{V}$ that contain the angle of the slipping contact α are rather smaller than the other terms — see graph — and therefore the correlation correction constitutes only a minor contribution.

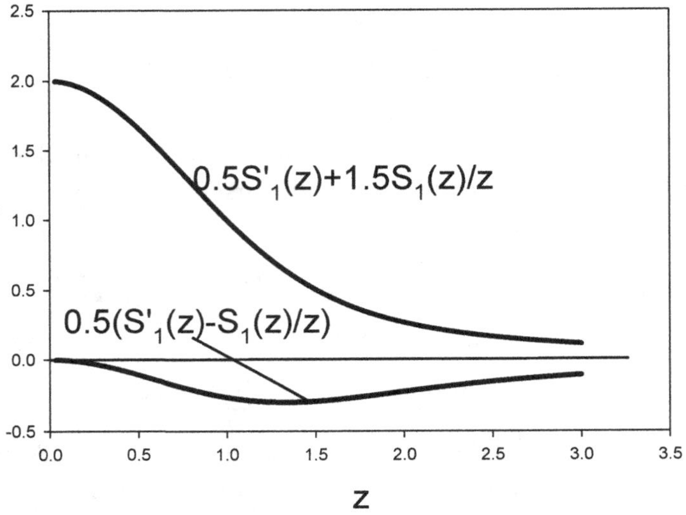

Figure 8.6. $S_1(z)$, $\frac{1}{2}S_1'(z)+\frac{3}{2z}S_1(z)$ and $\frac{1}{2}\left(S_1'(z)-\frac{1}{z}S_1(z)\right)$ as a function of z.

With the form for the strain fluctuations available it becomes possible to make an estimate of the stress increment as a function of the components of the mean strain increment. Naturally, the outcome depends on input parameters such as the number of contacts, the anisotropy, *etc*. The result is in part the same as the mean-field estimate, so all that needs to be reported here is the addition due to the displacement fluctuations as calculated above. There are two parts to this. The first is the part due the actual displacement fluctuations; this part is labelled A. The second contribution comes about as a result of the spin fluctuations; these contributions are labelled B.

The first extra term is as follows (note that averaging over the four directions has been carried out, so the angle α here refers to the first quadrant).

$$\frac{2vN_c}{N_V k\overline{c}^2}\left[\left(\frac{\hat{a}}{\overline{c}}\right)S_1\left(\frac{\overline{c}}{\hat{a}}\right)\right]^{-1}\sigma_{11}^{(slip-A)}$$

$$=\frac{1}{4}\sin 4\alpha\left[\mu_s\left(1-\sin^2 2\alpha\right)\left(\overline{e}_{11}-\overline{e}_{22}\right)+\frac{1}{2}\mu_s^2\sin 2\alpha\left(\overline{e}_{11}+\overline{e}_{22}\right)\right]$$

$$+\frac{1}{4}\left[\left(1-\mu_s^2\right)\sin^4 2\alpha-\left(2-\mu_s^2\right)\sin^2 2\alpha\right]\left(\overline{e}_{11}-\overline{e}_{22}\right)$$

$$+\frac{1}{4}\mu_s\left(2\sin 2\alpha-\sin^3 2\alpha\right)\left(\overline{e}_{11}+\overline{e}_{22}\right)$$

$$\frac{2vN_c}{N_V k\overline{c}^2}\left[\left(\frac{\hat{a}}{\overline{c}}\right)S_1\left(\frac{\overline{c}}{\hat{a}}\right)\right]^{-1}\sigma_{22}^{(slip-A)}$$

$$=\frac{1}{4}\sin 4\alpha\left[\mu_s\left(\sin^2 2\alpha-1\right)\left(\overline{e}_{11}-\overline{e}_{22}\right)-\frac{1}{2}\mu_s^2\sin 2\alpha\left(\overline{e}_{11}+\overline{e}_{22}\right)\right]$$

$$+\frac{1}{4}\left[\left(\mu_s^2-1\right)\sin^4 2\alpha+\left(2-\mu_s^2\right)\sin^2 2\alpha\right]\left(\overline{e}_{11}-\overline{e}_{22}\right)$$

$$+\frac{1}{4}\mu_s\left(\sin^3 2\alpha-2\sin 2\alpha\right)\left(\overline{e}_{11}+\overline{e}_{22}\right)$$

Reporting the extra stress increment due to the spin fluctuation, insofar as it is proportional to the components of **V**, it is convenient to introduce the shorthand $T_1=\frac{1}{2}S_1'\left(\frac{c}{\hat{a}}\right)+\frac{\hat{a}}{2c}S_1\left(\frac{c}{\hat{a}}\right); T_2=\frac{1}{2}\left(S_1'\left(\frac{c}{\hat{a}}\right)-\frac{\hat{a}}{c}S_1\left(\frac{c}{\hat{a}}\right)\right)$. The extra diagonal stress increments then turn out to be entirely proportional to T_1+T_2.

$$\frac{2vN_c}{N_V k\overline{c}^2\left(T_1+T_2\right)}\sigma_{11}^{(slip-B)}$$

$$=\frac{N_c a_p}{16\left(N_c-2\right)}\left(\overline{e}_{11}-\overline{e}_{22}\right)\sin 2\alpha\left(\sin 4\alpha-\mu_s\cos 4\alpha-\mu_s\sin 2\alpha\right)$$

$$-\frac{1}{16\left(N_c-2\right)}\left(\overline{e}_{11}-\overline{e}_{22}\right)\left(\mu_s\sin 8\alpha-\left(1-\mu_s^2\right)\sin^2 4\alpha\right)$$

$$-\frac{\mu_s}{16\left(N_c-2\right)}\left(\overline{e}_{11}+\overline{e}_{22}\right)\left(2\left(\cos 4\alpha+1\right)\sin 2\alpha+a_p N_c\sin 4\alpha\right)$$

$$\frac{2vN_c}{N_V k\overline{c}^2\left(T_1+T_2\right)}\sigma_{22}^{(slip-B)}$$
$$=\frac{N_c a_p}{16\left(N_c-2\right)}\left(\overline{e}_{11}-\overline{e}_{22}\right)\sin 2\alpha\left(\mu_s\cos 4\alpha-\sin 4\alpha+\mu_s\sin 2\alpha\right)$$
$$+\frac{1}{16\left(N_c-2\right)}\left(\overline{e}_{11}-\overline{e}_{22}\right)\left(\mu_s\sin 8\alpha-\left(1-\mu_s^2\right)\sin^2 4\alpha\right)$$
$$+\frac{\mu_s}{16\left(N_c-2\right)}\left(\overline{e}_{11}+\overline{e}_{22}\right)\left(2\left(\cos 4\alpha+1\right)\sin 2\alpha+a_p N_c\sin 4\alpha\right)$$

In practice, for a reasonable set of system parameters these B-contributions are negligible, compared to the A-contributions. For the shear stress then, it is sufficient to report the result of the A-contributions.

$$\frac{2vN_c}{N_V k\overline{c}^2}\left[\left(\frac{\overline{a}}{\overline{c}}\right)S_1\left(\frac{\overline{c}}{\overline{a}}\right)\right]^{-1}\sigma_{12}^{(slip-A)}$$
$$=\frac{1}{4}\sin 4\alpha\left[\left(1-\mu_s^2\right)\left(\sin 2\alpha-\frac{1}{2}\sin 4\alpha\right)+\mu_s\left(1-\sin^2 2\alpha\right)\right]\overline{e}_{12}$$
$$+\frac{1}{2}\mu_s\sin 2\alpha\left(2\sin^2 2\alpha-1\right)\overline{e}_{12}+\frac{1}{4}\sin 4\alpha\left(\sin\alpha-\mu_s\right)\overline{\eta}$$
$$+\frac{1}{4}\sin 2\alpha\left(\mu_s\left(1+\sin^2 2\alpha\right)+\sin 2\alpha\right)\overline{\eta}$$

The mean spin, $\overline{\eta}$, is determined either by requiring the spin fluctuations to average to zero, or simply by requiring stress symmetry.

$$\frac{\overline{\eta}}{\overline{e}_{12}}=\frac{\sin 2\alpha\left(2\mu_s\sin^2 2\alpha+\frac{1}{2}\left(1-\mu_s^2\right)\sin 4\alpha-\mu_s\right)N_V\left(\frac{\overline{a}}{\overline{c}}\right)S_1\left(\frac{\overline{c}}{\overline{a}}\right)}{\sin 2\alpha\left(\mu_s\cos 2\alpha-\sin 2\alpha\right)N_V\left(\frac{\overline{a}}{\overline{c}}\right)S_1\left(\frac{\overline{c}}{\overline{a}}\right)+N_c\left(N_c-1\right)}$$
$$+\frac{N_c\left(\cos 2\alpha+\mu_s-\frac{1}{2}a_p N_c\right)}{\sin 2\alpha\left(\mu_s\cos 2\alpha-\sin 2\alpha\right)N_V\left(\frac{\overline{a}}{\overline{c}}\right)S_1\left(\frac{\overline{c}}{\overline{a}}\right)+N_c\left(N_c-1\right)}$$

The non-dimensional determinant Δ and the dilatancy ratio δ have been plotted as a function of the angle of the slipping contact when there is one of these per particle and an average has been calculated over the four conjugate directions. It is seen that, with the parameter choice in the example of the graph that for contacts oriented towards the mean slipping angle α_s, the determinant vanishes and that the dilatancy peaks.

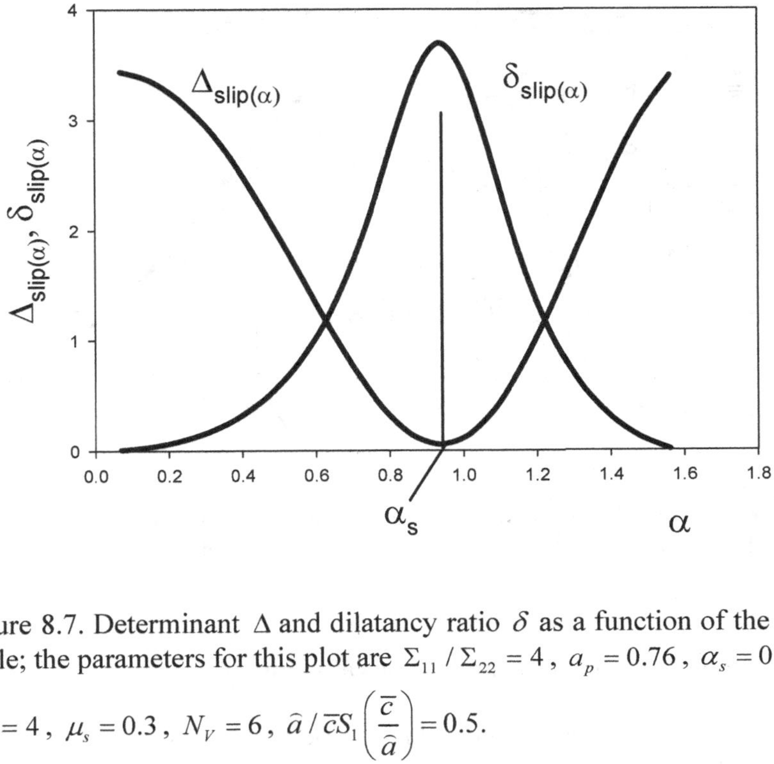

Figure 8.7. Determinant Δ and dilatancy ratio δ as a function of the slip angle; the parameters for this plot are $\Sigma_{11}/\Sigma_{22} = 4$, $a_p = 0.76$, $\alpha_s = 0.93$, $N_c = 4$, $\mu_s = 0.3$, $N_V = 6$, $\hat{a}/\overline{c}S_1\left(\dfrac{\overline{c}}{\hat{a}}\right) = 0.5$.

This is very similar to the features of the mean-field theory, only more pronounced.

The incremental shear stress ratio has also been evaluated and is depicted in Fig. 8.8. It is seen that, due to the strain increment fluctuations associated with the slipping contacts, the incremental shear modulus is substantially reduced, compared with the all-stick case. It is also observed that the value is rather smaller than the mean-field case. All this points to the characteristic features of granular assemblies with frictional effects and as such gives a refinement of the mean-field theory.

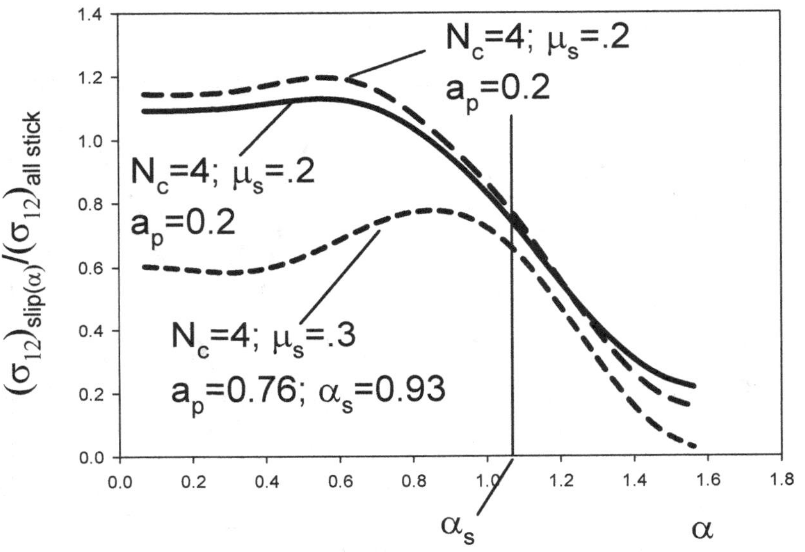

Figure 8.8. Shear stress ratio for an assembly with one slipping contact per particles to an all stick assembly. The same parameters as for Fig. 8.7 have been used.

8.6 Fabric heterogeneity and other refinements

There are some other refinements that can be considered. Note though that every refinement requires extra parameters and by introducing them the complexity of the modelling increases. Nevertheless, it is good to make a list of relevant mechanisms and show where the current — very simple — model is deficient.

1. So far it has been assumed that all particles have one slipping contact. This could be amended, especially taking into account that there will be a fraction of particles with no slipping contacts at all.

2. A certain amount of fabric heterogeneity could be introduced (as was done for the mean-field investigation, Section 8.4). This is a complication, as of course some fabric heterogeneity already follows from that associated with the sliding contacts.

3. Not all slipping contacts need to be bunched in the direction of α_s. Rather, a distribution of sliding contacts in the vicinity of α_s could be considered. Noting the sharpness of the peaks in the graphs of the dilatancy and the determinant, averaging over a region around the mean slip angle would be appropriate. Below it will be argued that this refinement is equivalent to the refinement number 2.

These three effects — while they would operate simultaneously — can be investigated separately to come to an understanding of how each affects the outcome of the calculation.

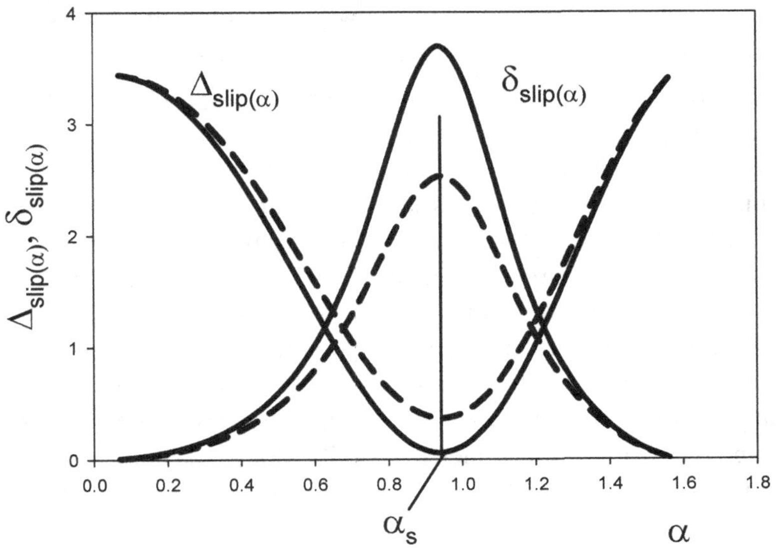

Figure 8.9. The same graph as Fig. 8.7, but with the addition of $\psi_s = 0.1$ as dashed lines.

To investigate the effect of not all particles having one slipping contact, refinement number 1, simply average the slipping model calculation above with the no-slide model. If there is a fraction of ψ_s no-slide contacts and $(1-\psi_s)$ sliders then the graph of Fig. 8.7 changes. This is depicted in Fig. 8.9 for a value of $\psi_s = 0.1$, but all other parameters kept the same as in Fig. 8.7.

It is seen that the determinant is larger; therefore, the assembly so-described would be further away from peak stress. Alternatively, it could be reasoned that this state requires a higher stress ratio (resulting in a higher anisotropy and marginally higher dilatancy).

Refinement 2, looks at the effect of fabric heterogeneity. The simplest way of doing this is by introducing a small value and thereby ignoring any interaction between the fabric variations due to sliding. However, it is noted that by assessing the displacement fluctuations due to sliding a slight overestimate is achieved, as the average interaction for a contact that slides is not the full 'stick' contact, but a slightly smaller value because one quarter of contacts in this direction slide. This effect is introduced as well as introducing a value for the contact point variations $\overline{(P')^2} / \overline{p}^2$. Just to see what this leads to, this parameter is set to a small value of 0.05 and the same graph as before can be made.

The result is shown in Fig. 8.10. The effect of the mean value of the no-slip interaction being somewhat smaller than the no-slip interaction itself lifts the determinant line slightly, while the effect of fabric fluctuations in the number of contacts pulls it down. The dilatancy is affected in the other direction: it is increased by fabric heterogeneity, though it depends on exactly how the parameters are chosen. It is noted that the two refinements 1 and 2 have opposite effects. Therefore, distinguishing between the two may be quite hard and the best thing is to combine the two in one parameter (a similar thing was done in the mean-field theory).

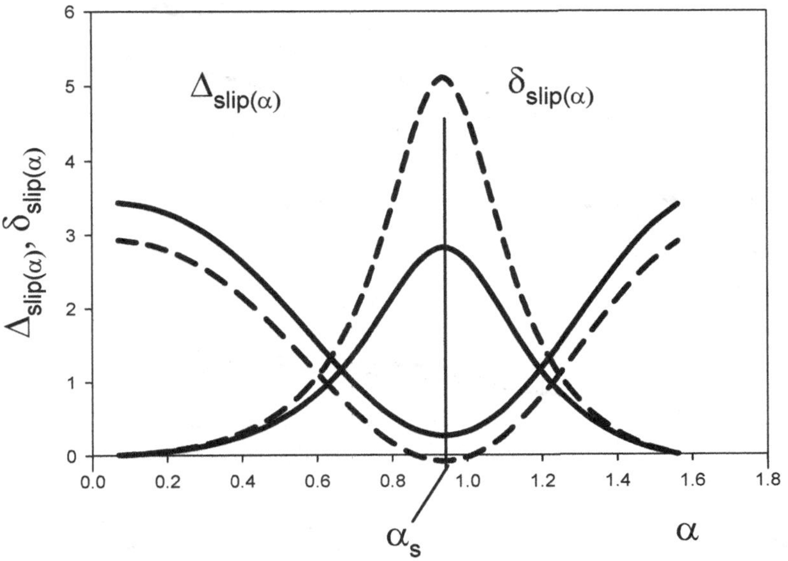

Figure 8.10. Same graph as Fig. 8.7, but added the two effects of fabric heterogeneity. Solid line: mean value of no-slip interaction $3/4$ of the value in Fig. 8.7. Dashed line: introduction of $\overline{(P')^2} / \overline{p}^2 = 0.05$.

The third refinement is the result of non-homogeneity in the packing properties, which result in local stress variations. These may lead to the direction of slipping angle being different from position to position. In refinement number 2 fabric heterogeneity has already been accounted for. However, the deformation fluctuation associated with fabric variations has not been coupled to variations in the direction of the slipping angle. The coupling would have only a small effect in the context of an assembly in which the sliding contacts are not dominant. It was shown in Section 7.6 that the dominant effect associated with fabric heterogeneity is the variation in the number of contacts. Thus, any other effects, such as contact point distribution fluctuations that are not aligned with the mean axis of anisotropy, are of lesser importance. As the parameter $\overline{(P')^2} / \overline{p}^2$ already comprises an amalgam of influences, it is not profitable to do a much more in-depth analysis.

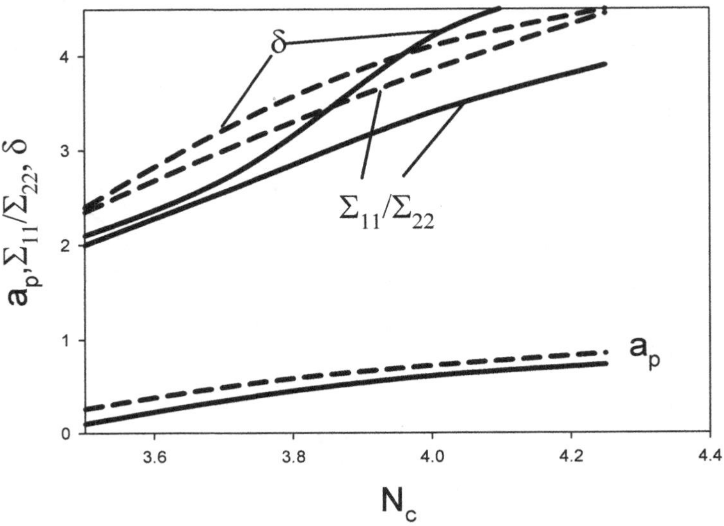

Figure 8.11. The anisotropy parameter, the stress ratio and the dilatancy parameter as a function of the mean number of contacts per particle. The solid lines correspond to a heterogeneity parameter of $H = 0.1$ and the dashed lines to $H = 0.05$. All calculations done at peak stress ratio when $\Delta = 0$.

It is now possible to construct a similar graph to Fig. 8.5, taking account of heterogeneity and the various refinements. This is reported in Fig. 8.11.

Qualitatively, the results are not all that different from the mean strain theory. For the lower end of the number of contacts the results are strikingly similar to experimental findings, for example as reported by [Konishi, 1978].

8.7 The evolution of an assembly in a biaxial cell test

A narrative for the evolution of an assembly of particles with a frictional interaction under deviatoric loading has been put forward by various authors: for example, [Cundall *et al.*, 1982], [Thornton and Antony, 1998]. Taking the biaxial test referred to in Section 1.1 the following scenario is plausible. The initial state is densely packed and statistically isotropic. This state is first isotropically compressed, then stressed further in one direction while the stress in the other direction is kept constant. As this deviatoric process progresses more contacts are made in the major principal stress direction while contacts are lost in the minor principal stress direction (more generally, the evolution of the contact point distribution tends to follow the stress ellipse). For geometrical reasons the number of contacts that can be made has a limited availability and the number of contacts that can be lost can in principle go on until there are very few left. Therefore, the total number of contacts is steadily reduced as the test progresses. At the same time the force ratio at contacts increases in the directions referred to in Section 8.4.2. In Section 8.3 the isotropic point of the stress-strain curve was discussed, while the peak-stress-ratio point was analysed in Section 8.4. Beyond this point rupture layer formation takes place and the problem will depend critically on the specification of the boundary conditions. The evolution from the isotropic point to the peak point could in principle be modelled with rate equations. However, a large number of parameters would have to be supplied (all of which amalgams representing various simultaneously operating effects).

In this section one more effect is considered that is relevant near peak stress ratio while the number of contacts decreases. This effect relates to Fig. 6.2. It is seen that when the number of contacts decreases, while the variability remains constant, the distance ratio $\overline{c}/\widehat{a}$ also increases, which results in a decrease in the function $S_1(\overline{c}/\widehat{a})/(\widehat{a}/\overline{c})$, as follows from Fig. 6.1. The result is that the effects of heterogeneity, both due to sliding and to fabric fluctuations become less intense and the value of the heterogeneity parameter $\widehat{H}$ (which is proportional to the S-function) goes down. In other words, the medium moves more towards a mean-field approximation, though some non-homogeneity is always present of course. This would happen close to the peak stress ratio point, when the determinant Δ is already small and the stress ratio will not change much.

When this occurs — consulting Fig. 8.5 — the dilatancy remains fairly constant. This has also been observed in experiments. In fact it is one of the key features of such tests.

While the analysis has been carried out in two dimensions, the physics that underlies the experimentally observed effects is adequately elucidated. The principal elements of the theory are summarised: (1) the treatment of sliding friction as a constraint, (2) the introduction of anisotropy as an integral part of the analysis and (3) the necessary consideration of fabric heterogeneity and a calculus to assess its impact, including the spatial spreading that gives rise to the S-functions. Insight in the latter aspect is enhanced by the analysis of connectivity in a granular medium (Section 6.6). All these aspects need to work together to acquire an understanding of the mechanisms that take place.

References

Cundall, P.A., Drescher, A. And Strack, O.D.L. (1982) Numerical experiments on granular assemblies; Measurements and observations. *Proc. IUTAM Conference on Deformation and Failure of Granular Materials*, Delft (Vermeer, P.A. and Luger, H.J. *eds.*) Rotterdam: Balkema.

Konishi, J. (1978) Microscopic model studies on the mechanical behaviour of granular material. *Proc. US-Japan Seminar on Continuum Mechanical and Statistical Approaches in the Mechanics of Granular Materials, Sendai, Japan* (eds. S.C. Cowin and M. Satake) pp. 27–45. Gakujutsu Bunken Fukya-Kai, Tokyo.

Thornton, C. and Antony, S.J. (1998) Quasi-static deformation of particulate media. *Phil. Trans. R. Soc. Lond. A* **356** 2763–2782.

Appendix A

Mathematical Appendix

A.1 Isotropic tensors

Literature: [Jeffreys, 1931].

A.1.1 *Isotropic 2-tensor*

The identity is δ_{ij}; its inverse is also δ_{ij}.

A.1.2 *Isotropic 4-tensor*

The identity of rank-4 tensors is such that a rank-4 tensor A_{ijpq}, which connects two symmetric rank-2 tensors has the inverse A^{-1}_{pqab}

$$A_{ijpq}A^{-1}_{pqab} = \frac{1}{2}\left(\delta_{ai}\delta_{bj} + \delta_{aj}\delta_{bi}\right)$$

The isotropic tensor of rank 4 is

$$\alpha\delta_{ab}\delta_{pq} + \beta\left(\delta_{ap}\delta_{bq} + \delta_{aq}\delta_{bp}\right)$$

Its inverse is

$$\xi\delta_{pq}\delta_{ij} + \psi\left(\delta_{ip}\delta_{jq} + \delta_{iq}\delta_{jp}\right) \text{ with } \psi = \frac{1}{4\beta} \text{ and } \xi = -\frac{\alpha}{4\beta(\alpha+\beta)} \text{ in 2-d}$$

$$\text{and } \xi = -\frac{\alpha}{2\beta(3\alpha+2\beta)} \text{ in 3-d.}$$

A.2 Integrals of strings of unit vectors

$$I^{(0)} = \int_0^{2\pi} d\phi = 2\pi$$

$$I^{(2)}_{ij} = \int_0^{2\pi} n_i(\phi)n_j(\phi)d\phi = \pi\delta_{ij}$$

$$I^{(4)}_{ijk\ell} = \int_0^{2\pi} n_i(\phi)n_j(\phi)n_k(\phi)n_\ell(\phi)d\phi = \frac{\pi}{4}\left(\delta_{ij}\delta_{k\ell} + \delta_{ik}\delta_{j\ell} + \delta_{i\ell}\delta_{kj}\right)$$

The inverse of the latter is the solution of the equation $I^{(4)}_{ijk\ell}\left(\mathbf{I}^{(4)}\right)^{-1}_{k\ell pq} = \tfrac{1}{2}\left(\delta_{ip}\delta_{jq} + \delta_{iq}\delta_{jp}\right)$ and

$$\left(\mathbf{I}^{(4)}\right)^{-1}_{k\ell pq} = \frac{1}{\pi}\left(-\frac{1}{2}\delta_{pq}\delta_{k\ell} + \delta_{pk}\delta_{q\ell} + \delta_{p\ell}\delta_{qk}\right)$$

In three dimensions the corresponding expressions are

$$I^{(0)} = \int_{unit\,sphere} d\Omega = 4\pi\,;\; I^{(2)}_{ij} = \int_{unit\,sphere} n_i n_j d\Omega = \frac{4\pi}{3}\delta_{ij}$$

and $$I^{(4)}_{ijk\ell} = \int_{unit\,sphere} n_i n_j n_k n_\ell d\Omega = \frac{4\pi}{15}\left(\delta_{ij}\delta_{k\ell} + \delta_{ik}\delta_{j\ell} + \delta_{i\ell}\delta_{jk}\right)$$

$$\left(\mathbf{I}^{(4)}\right)^{-1}_{k\ell pq} = \frac{15}{16\pi}\left(-\frac{2}{5}\delta_{pq}\delta_{k\ell} + \delta_{pk}\delta_{q\ell} + \delta_{p\ell}\delta_{qk}\right)$$

The front factors $2\pi, \pi, \pi/4$ in 2-d and $4\pi, 4\pi/3, 4\pi/15$ in 3-d are easily gathered in a coefficient

$$\frac{2(d-1)\pi}{(n+d-2)!!},$$

where $(2m)!! = 2.4....(2m)$; $(2m-1)!! = 1.3.5...(2m-1)$, m an integer. (see [Abramowitz and Stegun, 1965], Section 6.1.49).

A very useful coefficient is

$$\aleph^{(n,d)} = \frac{1}{(n+d-2)!!}.$$

A.3 Elastic constants

Conversion of the isotropic elastic constants in two and three dimensions. Top line: 3-D; bottom line 2-D.

λ and μ: Lamé constants

E : Young's modulus, ν : contraction coefficient (Poisson's ratio)

K : bulk modulus, G : shear modulus

	E,ν	λ,μ	K,G
E	E	$\frac{\mu(3\lambda+2\mu)}{\lambda+\mu}$; $\frac{4\mu(\lambda+\mu)}{\lambda+2\mu}$	$\frac{9GK}{G+3K}$; $\frac{4GK}{G+K}$
ν	ν	$\frac{\lambda}{2\lambda+2\mu}$; $\frac{\lambda}{\lambda+2\mu}$	$\frac{3K-2G}{2(G+3K)}$; $\frac{K-G}{K+G}$
λ	$\frac{E\nu}{(\nu+1)(1-2\nu)}$; $\frac{E\nu}{(\nu+1)(1-\nu)}$	λ	$K-\frac{2}{3}G$; $K-G$
μ	$\frac{E}{2(1+\nu)}$	μ	G
K	$\frac{E}{3(1-2\nu)}$; $\frac{E}{1-\nu}$	$\lambda+\frac{2}{3}\mu$; $\lambda+\mu$	K
G	$\frac{E}{2(1+\nu)}$	μ	G

A.4 Fourier transforms and harmonic density

Fourier transforms are a powerful tool to solve differential equations. In this section the basic theory is explored, which is required for the characterisation of the fluctuations. To begin with a one-dimensional approach is taken, in which functions depend on one variable, the time t say. In the development below it makes sense to keep the integral sign with its boundaries together with the integration variable.

The appropriate tool for describing fluctuating physical phenomena is the *auto-correlation function*. For a fluctuating function of time $z(t)$ (which is zero on average) it is defined as

$$\phi_z(t) = \lim_{T\to\infty} \frac{1}{T} \int_0^T d\xi\, z(\xi) z(t+\xi)$$

It is seen that the expectation value $\langle z^2 \rangle$ is just equal to $\phi_z(0)$.

The Fourier transform $\hat{z}(\omega)$ of the function $z(t)$ is defined as

$$\hat{z}(\omega) = \int_{-\infty}^{\infty} dt\, z(t) e^{-i\omega t}$$

The inverse transform is

$$z(t) = \frac{1}{2\pi} \int_{-\infty}^{\infty} d\omega\, \hat{z}(\omega) e^{i\omega t},$$

which makes the delta function (the 'identity')

$$\delta(t) = \frac{1}{2\pi} \int_{-\infty}^{\infty} d\omega\, e^{i\omega t}$$

The latter has meaning only in the context of another function, as follows

$$z(t) = \int_{-\infty}^{\infty} d\xi\, z(\xi) \delta(t-\xi)$$

In physical processes the infinite integral boundaries do not make sense and therefore a modified definition is employed, the truncated Fourier transform

$$\hat{z}_\tau(\omega) = \int_0^\tau dt\, z(t) e^{-i\omega t}$$

Now calculate the inverse Fourier transform of the quantity $\hat{S}_z \equiv \hat{z}_\tau(\omega)\hat{z}_\tau(-\omega)/\tau$

$$S_z(t) = \frac{1}{2\pi\tau}\int_{-\infty}^{\infty} d\omega\, e^{i\omega t}\int_0^\tau d\lambda\, z(\lambda)e^{-i\omega\lambda}\int_0^\tau d\mu\, z(\mu)e^{i\omega\mu}$$

The order of the integrals may be interchanged and therefore

$$\begin{aligned} S_z(t) &= \frac{1}{2\pi\tau}\int_0^\tau d\lambda\int_0^\tau d\mu\int_{-\infty}^{\infty} d\omega e^{i\omega(t-\lambda+\mu)}\, z(\lambda)z(\mu) \\ &= \frac{1}{\tau}\int_0^\tau d\lambda\int_0^\tau d\mu\, \delta(t-\lambda+\mu)z(\lambda)z(\mu) \end{aligned}$$

Integrating over λ (using the properties of the delta function) gives

$$S_z(t) = \frac{1}{\tau}\int_0^{\tau-t} d\mu\, z(t+\mu)z(\mu)$$

For a time record τ that is much longer than the correlation time the upper boundary may be replaced by τ and it is seen that in the limit $\tau \to \infty$ the inverse Fourier transform $S_z(t)$ is just equal to the correlation function $\phi_z(t)$. This is the famous Wiener–Khinchin theorem. The quantity $\hat{S}_z(\omega)$ is called the spectral intensity or harmonic (spectral) density.

In two or three dimensions the formulas are easily extended, by taking the integrals over multiple variables. The Fourier frequency ω then becomes a vector. In a spatial setting this vector is called the wave number **k**.

A.5 Bessel functions

Literature [Abramowitz and Stegun, 1965].

Bessel functions are very useful in the evaluation of problems that have cylinder symmetry. There are various types. The best-known is the family of 'ordinary' Bessel functions, or *Bessel functions of the first kind.* They can either be defined as a series expansion

$$J_\nu(z)=\left(\frac{1}{2}z\right)^\nu\sum_{k=0}^{\infty}\frac{\left(-\frac{1}{2}z^2\right)^k}{k!\Gamma(\nu+k+1)}$$

Or as an integral

$$J_\nu(z)=\frac{\left(\frac{1}{2}z\right)^\nu}{\sqrt{\pi}\Gamma\left(\nu+\frac{1}{2}\right)}\int_0^\pi\cos(z\cos\theta)\sin^{2\nu}\theta d\theta,$$

where ν denotes the order of the Bessel function, indicating which member of the family is meant.

The other type of cylinder functions that are useful are the *Modified Bessel Functions.*

$$I_\nu(z)=\left(\frac{1}{2}z\right)^\nu\sum_{k=0}^{\infty}\frac{\left(\frac{1}{4}z^2\right)^k}{k!\Gamma(\nu+k+1)}$$

Or

$$I_\nu(z)=\frac{\left(\frac{1}{2}z\right)^\nu}{\sqrt{\pi}\Gamma\left(\nu+\frac{1}{2}\right)}\int_0^\pi\exp(\pm z\cos\theta)\sin^{2\nu}\theta d\theta$$

The Bessel functions have been studied extensively. There are all manner of interesting relations between them. Many of these can be found in [Abramowitch and Stegun, 1965].

Special cases of the half-integer Modified Bessel Functions are

$$I_{\frac{1}{2}}(z)=\frac{\sinh z}{\sqrt{\frac{1}{2}\pi z}};\ I_{\frac{3}{2}}(z)=\frac{1}{\sqrt{\frac{1}{2}\pi}}\left(-\frac{\sinh z}{z^{3/2}}+\frac{\cosh z}{z^{1/2}}\right)$$

A.6 Various integrals

A.6.1 *Integrals involving Bessel functions*

Literature [Gradshteyn and Ryzhik, 1965], abbreviated as GR.

$$\int_0^{2\pi} \cos(z\cos\theta)\,d\theta = 2\pi J_0(z)$$

$$i^{-n}\int_0^{\pi} \exp(iz\cos\theta)\cos(nz)\,d\theta = \pi J_n(z) \quad \text{see also GR p 402; 3.715.14}$$

In particular with $\hat{n}_1 = \cos\psi;\ \hat{n}_2 = \sin\psi;\ m_1 = \cos\varphi;\ m_2 = \sin\varphi$

$$\int_0^{2\pi} d\psi \sin\left[kx\cos(\psi-\varphi)\right]\hat{n}_j =$$

$$\left\{\begin{array}{l} \int_0^{2\pi} d\alpha \sin[kx\cos\alpha]\cos(\alpha+\varphi) = \cos\varphi\int_0^{2\pi} d\alpha\sin[kx\cos\alpha]\cos\alpha \\ \int_0^{2\pi} d\alpha \sin[kx\cos\alpha]\sin(\alpha+\varphi) = \sin\varphi\int_0^{2\pi} d\alpha\sin[kx\cos\alpha]\cos\alpha \end{array}\right\}$$

$$= 2\pi J_1(kx)m_j$$

$$\int_0^{2\pi} d\psi\sin\left[kx\cos(\psi+\varphi)\right]\hat{n}_j\hat{n}_a\hat{n}_i = \{111,112,122,222\} =$$

$$2\pi\left\{\frac{3}{4}J_1(kx)\cos\varphi - J_3(kx)\cos\varphi\left(\cos^2\varphi-\frac{3}{4}\right),\right.$$

$$\frac{1}{4}J_1(kx)\sin\varphi - J_3(kx)\sin\varphi\left(\cos^2\varphi-\frac{1}{4}\right),$$

$$\frac{1}{4}J_1(kx)\cos\varphi - J_3(kx)\cos\varphi\left(\frac{3}{4}-\cos^2\varphi\right),$$

$$\left.\frac{3}{4}J_1(kx)\sin\varphi - J_3(kx)\sin\varphi\left(\frac{1}{4}-\cos^2\varphi\right)\right\} =$$

$$\frac{\pi}{2}J_1(kx)\left(m_i\delta_{aj}+m_j\delta_{ai}+m_a\delta_{ij}\right)$$

$$+\frac{\pi}{2}J_3(kx)\left(m_i\delta_{aj}+m_j\delta_{ai}+m_a\delta_{ij}\right) - 2\pi J_3(kx)m_im_jm_a$$

$$\int_0^\infty yJ_0(ky)\exp\left(-\frac{y^2}{a^2}\right)dy=\frac{a^2}{2}\exp\left(-\frac{1}{4}a^2k^2\right) \quad \text{see GR p717; 6.631.4}$$

$$S_1\left(\frac{x}{a}\right)\equiv a\int_0^\infty dk\exp\left(-\frac{1}{4}a^2k^2\right)J_1(kx)=\sqrt{\pi}\exp\left(-\frac{x^2}{2a^2}\right)I_{\frac{1}{2}}\left(\frac{x^2}{2a^2}\right)$$

$$S_3\left(\frac{x}{a}\right)\equiv a\int_0^\infty dk\exp\left(-\frac{1}{4}a^2k^2\right)J_3(kx)=\sqrt{\pi}\exp\left(-\frac{x^2}{2a^2}\right)I_{\frac{3}{2}}\left(\frac{x^2}{2a^2}\right)$$

see GR p710; 6.618.1.

These functions are used in Equation (6.4).

A.6.2 *Integrals with confluent hypergeometric functions*

The confluent hypergeometric function can be defined as[a]

$$\Phi(a;b;z)=\frac{\Gamma(b)}{\Gamma(a)\Gamma(b-a)}\int_0^1 e^{zt}t^{a-1}(1-t)^{b-a-1}\,dt$$

Other definitions (for example, a series expansion) are listed in [Abramowitz and Stegun, 1965, Chapter 13].

The special case that is needed here is for $a=\frac{1}{2}$; $b=\frac{3}{2}$

$$\Phi\left(\frac{1}{2};\frac{3}{2};z\right)=-\frac{i\sqrt{\pi}}{2\sqrt{z}}erf\left(i\sqrt{z}\right)$$

It arises from the integral [Gradshteyn and Ryzhik 3.896.4]

$$\int_0^\infty dk\sin(ky)\exp\left(-\frac{1}{4}a^2k^2\right)=\frac{2y}{a^2}\exp\left(-\frac{y^2}{a^2}\right)\Phi\left(\frac{1}{2};\frac{3}{2};\frac{y^2}{a^2}\right)$$

This expression has a series expansion in y/a

---

[a]The Gamma function $\Gamma(z)=\int_0^\infty t^{z-1}e^{-t}dt$; $\Gamma(n+1)=n\Gamma(n)=n!$; $\Gamma\left(\frac{1}{2}\right)=\sqrt{\pi}$.

$$\frac{1}{a}\left(2\frac{y}{a}+\frac{2}{3}\left(\frac{y}{a}\right)^2-\frac{9}{5}\left(\frac{y}{a}\right)^3-\frac{13}{21}\left(\frac{y}{a}\right)^4+\frac{437}{540}\left(\frac{y}{a}\right)^5+\ldots\right)$$

A.6.3 *Multiple integrals*

The integral $\frac{1}{(2\pi)^n}\int d^n k e^{i\mathbf{k}.\mathbf{x}}\frac{k_j k_b}{k^2}$ in two dimensions ($n=2$) diverges. In three dimensions integrate first over the component of $\mathbf{k}$ that does not appear in the subscripts j or b. Assume that that is the third component

$$\frac{1}{(2\pi)^3}\int d^3 k e^{i\mathbf{k}.\mathbf{x}}\frac{k_j k_b}{k^2}=\frac{1}{(2\pi)^3}\int d^2 k e^{i\mathbf{k}.\mathbf{x}}k_j k_b\int_{-\infty}^{\infty}dk_3 e^{ik_3 z}\frac{1}{\left(k^2+k_3^2\right)}$$

$$=\frac{2}{(2\pi)^3}\int d^2 k e^{i\mathbf{k}.\mathbf{x}}k_j k_b\int_0^{\infty}dk_3\cos(k_3 z)\frac{1}{\left(k^2+k_3^2\right)}$$

$$=\frac{1}{2(2\pi)^2}\int d^2 k e^{i\mathbf{k}.\mathbf{x}}k_j k_b\frac{1}{k}e^{-kz}$$

$$=\frac{-1}{2(2\pi)^2}\frac{\partial^2}{\partial x_j\partial x_b}\int d^2 k e^{i\mathbf{k}.\mathbf{x}}\frac{1}{k}e^{-kz}$$

$$=\frac{-1}{2(2\pi)^2}\frac{\partial^2}{\partial x_j\partial x_b}\int_0^{2\pi}d\varphi\int_0^{\infty}dk e^{ikx\cos(\varphi-\alpha)}e^{-kz}$$

$$=\frac{-1}{(2\pi)^2}\frac{\partial^2}{\partial x_j\partial x_b}\int_0^{\infty}dk e^{-kz}J_0(kx)=\frac{-1}{(2\pi)^2}\frac{\partial^2}{\partial x_j\partial x_b}\frac{1}{\sqrt{x^2+z^2}}$$

Here $z>0$; there is no problem for negative z, as it is seen immediately from the cosine that the answer is the same. For the differentiation and integration to be exchanged all the integrals have to exist, which is the case if both x and z are positive.

The integral $\frac{1}{(2\pi)^n}\int d^n k e^{i\mathbf{k}.\mathbf{x}}\frac{1}{k^4}k_a k_i k_j k_b$ in two dimensions again diverges. In three dimensions a finite result is obtained if first the calculation is done for a combination of subscripts that excludes one of the coordinates. Let this be the third coordinate.

$$\frac{1}{(2\pi)^3}\int d^3ke^{i\mathbf{k.x}}\frac{k_ak_bk_ik_j}{k^4}=\frac{1}{(2\pi)^3}\int d^2ke^{i\mathbf{k.x}}k_ak_bk_ik_j\int_{-\infty}^{\infty}dk_3e^{ik_3z}\frac{1}{\left(k^2+k_3^2\right)^2}$$

$$=\frac{2}{(2\pi)^3}\int d^2ke^{i\mathbf{k.x}}k_ak_bk_ik_j\int_0^{\infty}dk_3\cos(k_3z)\frac{1}{\left(k^2+k_3^2\right)^2}$$

$$=\frac{1}{4(2\pi)^2}\int d^2ke^{i\mathbf{k.x}}k_ak_bk_ik_j\frac{1}{k^3}(1+kz)e^{-kz}$$

Having established that this integral never diverges while both x and z are positive, avoiding having to deploy complicated combinations of Bessel functions, the evaluation is approached again as a derivative of the simpler integral. However, this one does diverge

$$\frac{1}{4(2\pi)^2}\int d^2ke^{i\mathbf{k.x}}k_ak_bk_ik_j\frac{1}{k^3}(1+kz)e^{-kz}$$

$$=\frac{1}{4(2\pi)^2}\frac{\partial^4}{\partial x_a\partial x_b\partial x_i\partial x_j}\int d^2ke^{i\mathbf{k.x}}\frac{1+kz}{k^3}e^{-kz}$$

Therefore, before the integral over k is performed two differentiations have to be executed to obtain a finite result.

$$\frac{1}{4(2\pi)^2}\frac{\partial^4}{\partial x_a\partial x_b\partial x_i\partial x_j}\int d^2ke^{i\mathbf{k.x}}\frac{1+kz}{k^3}e^{-kz}$$

$$=\frac{1}{2(2\pi)^2}\frac{\partial^4}{\partial x_a\partial x_b\partial x_i\partial x_j}\int_0^{\infty}dk\frac{1+kz}{k^2}e^{-kz}J_0(kx)$$

Now,

$$\frac{\partial^2 J_0(kx)}{\partial x_i \partial x_j} = \frac{\partial}{\partial x_i}\left[-J_1(kx)k\frac{\partial x}{\partial x_j}\right] =$$

$$-\left[J_0(kx) - \frac{1}{kx}J_1(kx)\right]k^2\frac{\partial x}{\partial x_j}\frac{\partial x}{\partial x_i} - J_1(kx)k\frac{\partial^2 x}{\partial x_i \partial x_j}$$

So, the integrals that have to be done are

$$\int_0^\infty dk(1+kz)e^{-kz}J_0(kx) = \frac{1}{\sqrt{x^2+z^2}} + \frac{z^2}{(x^2+z^2)^{3/2}}$$

$$\int_0^\infty dk(1+kz)e^{-kz}\frac{1}{k}J_1(kx) = \frac{x}{\sqrt{x^2+z^2}}$$

It follows that

$$\frac{1}{2(2\pi)^2}\frac{\partial^4}{\partial x_a \partial x_b \partial x_i \partial x_j}\int_0^\infty dk\frac{1+kz}{k^2}e^{-kz}J_0(kx)$$

$$= \frac{1}{2(2\pi)^2}\frac{\partial^2}{\partial x_a \partial x_b}\int_0^\infty dk\frac{1+kz}{k^2}e^{-kz}J_0(kx)$$

$$\frac{\partial}{\partial x_i}\left(J_1(kx)\frac{x_j}{x}\right)$$

$$= J_1(kx)\frac{\delta_{ji}}{x} - J_1(kx)\frac{x_j x_i}{x^2} + \frac{x_j x_i}{x^2}k\left(J_0(kx) - \frac{1}{kx}J_1(kx)\right)$$

$$\frac{1}{(2\pi)^2}\int d^2k e^{i\mathbf{k.x}}\frac{1}{k^4}k_a k_i k_j k_b =$$

$$\frac{1}{(2\pi)^2}\frac{\partial^4}{\partial x_a \partial x_b \partial x_i \partial x_j}\int_0^\infty dk \frac{1}{k^3}\int_0^{2\pi} d\varphi e^{ikx(\cos\alpha\cos\varphi+\sin\alpha\sin\varphi)} =$$

$$\frac{1}{2\pi}\frac{\partial^4}{\partial x_a \partial x_b \partial x_i \partial x_j}\int_0^\infty dk \frac{1}{k^3}J_0(kx) = \frac{-1}{2\pi}\frac{\partial^3}{\partial x_a \partial x_b \partial x_i}\int_0^\infty dk \frac{1}{k^2}J_1(kx)\frac{\partial x}{\partial x_j}$$

$$\frac{-1}{2\pi}\frac{\partial^2}{\partial x_a \partial x_b}\int_0^\infty dk \frac{1}{k^2}\frac{\partial}{\partial x_i}\left(J_1(kx)\frac{x_j}{x}\right)$$

$$= \frac{1}{2\pi}\frac{\partial}{\partial x_j}\left(\frac{1}{x}\frac{\partial x}{\partial x_b}\right) = \frac{1}{2\pi}\frac{\partial}{\partial x_j}\left(\frac{x_b}{x^2}\right) = \frac{1}{2\pi}\left(\frac{\delta_{jb}}{x^2} - 2\frac{x_b x_j}{x^4}\right)$$

$$\int_0^\infty dk_1 \frac{\cos(k_1 x_1)}{(k_1^2+k_2^2)^2} = \frac{\pi}{4k_2^3}(1+k_2 x_1)e^{-k_2 x_1} \text{ while both } k_2,\, x_1 > 0$$

Entirely analogous to the previous case one obtains

$$\frac{1}{(2\pi)^2}\int d^2k e^{i\mathbf{k.x}}\frac{1}{k^4}k_a k_i k_j k_b =$$

$$\frac{4}{(2\pi)^2}\frac{\partial^4}{\partial x_a \partial x_b \partial x_i \partial x_j}\int_0^\infty dk_1 \int_0^\infty dk_2 \frac{\cos(k_1 x_1)\cos(k_2 x_2)}{(k_1^2+k_2^2)^2} =$$

$$\frac{1}{4\pi}\frac{\partial^4}{\partial x_a \partial x_b \partial x_i \partial x_j}\int_0^\infty dk_2 \cos(k_2 x_2)\frac{1}{k_2^3}(1+k_2 x_1)e^{-k_2 x_1}$$

By inspecting symmetry relations the régime for all values of $x_{1,2}$ is easily determined.

The result is

$$\text{Subscripts } 1111 \quad -\frac{x_1^4 + 6x_1^2 x_2^2 - 3x_2^4}{4\pi r^6}$$

$$\text{Subscripts } 1122 \quad -\frac{x_1^4 - 6x_1^2 x_2^2 + x_2^4}{4\pi r^6}$$

Subscripts 2222 $\dfrac{3x_1^4 - 6x_1^2x_2^2 - x_2^4}{4\pi r^6}$

Subscripts 1112 $-\dfrac{|x_1||x_2|\left(x_1^2 - 3x_2^2\right)}{2\pi r^6}$

Subscripts 1222 $\dfrac{|x_1||x_2|\left(x_2^2 - 3x_1^2\right)}{2\pi r^6}$

References

Abramowitz, M. and Stegun, I.A. (1965) *Handbook of Mathematical Functions*. New York: Dover Publications.

Gradshteyn, I.S. and Ryzhik, I.M. (1965) *Table of Integrals Series and Products*. New York: Academic Press.

Jeffreys, H. (1931) *Cartesian Tensors*. Cambridge: Cambridge University Press.

Appendix B

List of Symbols and Notations

B.1 List of symbols

a	Particle radius
$\hat{a}$	Length parameter, describing the influence region of a fluctuation
a_p	Contact point anisotropy parameter
a_s	Simple shear parameter
A	Surface area
$\mathbf{A}$	Wave amplitude (inc polarisation)
A_{12}	Hamaker constant
$\mathbf{c}^{\bullet\bullet}$	Coordination vector
$\overline{c}$	Average magnitude of the coordination vector
$\mathbf{C}$	Compliance tensor
d	Dimension of the problem (either 2 or 3)
$d_{\bullet}$	grainsize below which •% of the weight of the sample is measured
$d_V^{\bullet}$	Distance from particle centre • to the nearest Voronoi boundary
$d_{\perp}$	Normal contact displacement increment
$d_{\parallel}$	Tangential contact displacement increment
$d_{\diamond}$	Intermediate contact displacement increment
$\mathbf{d}$	Deviatoric part of the contact distribution
$D_{\perp}$	Normal contact displacement
$D_{\parallel}$	Tangential contact displacement
D / Dt	Co-moving derivative
$E,\ E'$	Young's moduli of two particles in contact

$\mathbf{E}$	Strain tensor
$\mathbf{e}$	Strain increment tensor
e	Electron charge
f	Fraction
$\mathbf{f}$	Force increment
$f_\perp$	Normal contact force increment
$f_\parallel$	Tangential contact force increment
$f_\diamond$	Intermediate contact force increment
f_μ	Fraction of slipping contacts in an assembly
f_d	Factor $\frac{\pi}{4}$ in 2-d and $\frac{4\pi}{15}$ in 3-d
$\mathbf{F}$	Force vector
$\mathbf{F}$	Influence function fourth order tensor
$F_\perp$	Normal contact force
$F_\parallel$	Tangential contact force
$F_\diamond$	Intermediate contact force
$\mathbf{g}(\mathbf{x}\cdot\mathbf{n})$	Displacement field of a rupture layer with unit normal $\mathbf{n}$
$\mathbf{g}$	Acceleration due to gravity
$\mathbf{G}$	$\left\lvert\int_{-\infty}^{\infty} dx_3 \mathbf{F}\right\rvert$
G	Shear modulus
H	Separation between two particles
$\widehat{H}$	Heterogeneity parameter for an assembly with frictionally sliding contacts (Section 8.4.2)
J_n	Bessel function of order n
k	Contact interaction spring constant
$k^{\mu\nu}$	Contact interaction spring constant between particles μ and ν
k_B	Boltzmann's constant
$\begin{pmatrix} k_{\perp\perp} & k_{\perp\parallel} \\ k_{\parallel\perp} & k_{\parallel\parallel} \end{pmatrix}$	Normal and tangential contact stiffness matrix

$\begin{pmatrix} k_{\perp\perp} & 0 & \mu_s^{-1}k_{\parallel\Diamond} \\ \mu_s k_{\perp\perp} & 0 & k_{\parallel\Diamond} \\ k_{\Diamond\perp} & k_{\Diamond\parallel} & k_{\Diamond\Diamond} \end{pmatrix}$	Contact stiffness matrix in three dimensions
$\mathbf{k}$	Contact interaction tensor
$\mathbf{k}$	Fourier wave vector; magnitude $k = \lvert\mathbf{k}\rvert$
K	Bulk modulus
$L_1,\ L_2$	Length parameters
ℓ	Length of cylinder
$\ell(\Omega)$	Angular density of the population at the assembly
$\mathbf{m}$	Unit normal vector
n	Porosity
$n^{(0)}$	Bulk concentration of ions
$\mathbf{n}$	Unit normal vector
$\overline{\mathbf{n}}$	Unit normal vector, normal to $\mathbf{n}$
$\mathbf{n}_{\parallel}\ \mathbf{n}_{\Diamond}$	Unit normals in the direction of the tangential and intermediate force
N	Number of particles in an assembly
$N_c^{\bullet}$	Number of contacts of particle •
$N_{c,iso}$	Number of contacts per particle in the isostatic limit
p	Pressure
P'	Isotropic fluctuation in the contact distribution
$p^{\bullet}(\ )$	Angular contact distribution of particle •
$\mathbf{p}^{\bullet}$	Two-tensor denoting the symmetrical contact distribution of particle •
$p_0,\ p_1$	Influence parameters (inverse stiffnesses), Sections 6.5 and 7.4
$\overline{p}$	$\frac{1}{2}(\overline{p}_{11} + \overline{p}_{22})$
$\mathbf{q}^{\bullet}$	Vector denoting the asymmetry of the contact distribution of particle •
$P_1,\ P_2$	Principal pre-stresses
$\mathbf{P}$	Acoustic tensor

Q	$Q=\frac{3}{4}\left(\frac{1-\nu^2}{E}+\frac{1-\nu'^2}{E'}\right)$
$\mathbf{Q}(t)$	Time-dependent rigid coordinate rotation
$\mathbf{Q}(\alpha)$	Coordinate rotation over an angle α
$\mathbf{r}$	Rotation increment
R, R'	Radii of contacting particles
R	Radius of spherical (circular) assembly
R	Pre-stress ratio
$\hat{R}^{\bullet}$	Cell contact radius of particle •
$\mathbf{R}$	Local response tensor
S_1, S_3	Functions defined in Section
$\mathbf{S}$	Stiffness cross-correlation function
t	Time
$\mathbf{t}$	Traction vector increment
T	Absolute temperature
$\mathbf{T}$	Traction vector
$\hat{\mathbf{t}}$	Average tensor that gives the contact force increment
$\hat{\mathbf{T}}$	Average tensor that gives the contact force
$T_{1,2}$	$T_1=\frac{1}{2}S_1'\left(\frac{c}{\hat{a}}\right)+\frac{\hat{a}}{2c}S_1\left(\frac{c}{\hat{a}}\right);T_2=\frac{1}{2}\left(S_1'\left(\frac{c}{\hat{a}}\right)-\frac{\hat{a}}{c}S_1\left(\frac{c}{\hat{a}}\right)\right).$
$\mathbf{u}$	Displacement vector
$\mathbf{v}$	Velocity vector
v	Equivalent mean volume of one particle $v=V/N$
V	Volume
$V(H)$	Interactive potential
V_T	Total interactive potential
$\mathbf{V}$	Matrix form of the positional part of the first displacement derivative for a sliding contact (Section 8.5.3)
W	Work
$\mathbf{x}$, $\mathbf{y}$	Position vectors; magnitudes $x=\lvert\mathbf{x}\rvert$, $y=\lvert\mathbf{y}\rvert$
$\mathbf{X}$	Stiffness tensor
$\mathbf{X}^{mf}$	Mean-field stiffness tensor
y	Tangent of the rupture layer angle, squared
z	Non-dimensional parameter

Z	Valency of the ions
$\mathbf{Z}'$	Fluctuation source term of the differential equation (7.2)
$\boldsymbol{\alpha}$	Assembly-average displacement gradient
δ	Dilatancy ratio
$\boldsymbol{\delta}$	Kronecker delta tensor
ε	Electrical permittivity of the fluid
$\boldsymbol{\varepsilon}$	Levi-Civita tensor
ϕ	Solids volume fraction (solidosity)
φ	Angle in the two-dimensional plane
φ_{ne}	No-extension direction
$\phi_{d_V}, \tilde{\phi}_{d_V}$	Correlation function and normalised correlation function
κ	Reciprocal of the double layer thickness
$\lambda,\ \mu$	Lamé constants
$\boldsymbol{\lambda}$	Set of Lagrange multipliers
Δ	Non-dimensional outer determinant of the stiffness tensor
$\boldsymbol{\Lambda}^{\bullet}$	Tensor, giving the proportionality of the spin to the mean strain
μ_s	Friction coefficient
$\nu,\ \nu'$	Poisson's ratios of two particles in contact
χ	Phase of a wave
ψ_0	Surface potential
ρ	Mass density
$\boldsymbol{\sigma}$	Stress increment tensor
$\boldsymbol{\Sigma}$	Stress tensor
$\overset{\nabla}{\boldsymbol{\Sigma}}$	Jaumann stress-rate
$\boldsymbol{\xi}$	Material coordinate vector
$\boldsymbol{\Xi}$	Spin control tensor $\Xi_{in} = \varepsilon_{ijk}\varepsilon_{\ell mn}\overline{A}_{jlkm}$
ς	Material constant
Ω	Solid angle
$\Im$	Amount of energy per unit volume

B.2 List of notations

* Designation of components of a tensor in a rotated coordinated frame

Particles are numbered. The numbers are identified as a Greek superscript. For example $\mathbf{x}^{\mu}$ is the position vector of particle μ. Two Greek superscripts are used to denote a property of a particle pair. Examples, the coordination vector is $\mathbf{c}^{\mu\nu} = \mathbf{x}^{\nu} - \mathbf{x}^{\mu} = -\mathbf{c}^{\nu\mu}$, the unit normal $\mathbf{n}^{\mu\nu} = \mathbf{c}^{\mu\nu} / c^{\mu\nu} = -\mathbf{n}^{\nu\mu}$ and the contact force exerted by particle ν on particle μ is $\mathbf{F}^{\mu\nu} = -\mathbf{F}^{\nu\mu}$. Generally, pair-*vectors* change sign when the superscripts are interchanged. Second order tensors do not change sign, for example, the interaction tensor $\mathbf{K}^{\mu\nu} = \mathbf{K}^{\nu\mu}$.

B.2.1 *Structural sums*

Structural sums take the interaction and sum them over the Voronoi boundaries of a particle, weighed with components of branch vectors. They are called $\mathbf{A}^{\bullet}$. The lowest order ones are

$$A_{ij}^{\mu} = \sum_{\nu=1}^{N^{\mu}} K_{ij}^{\mu\nu} \; ; \; A_{ijk}^{\mu} = \sum_{\nu=1}^{N^{\mu}} K_{ij}^{\mu\nu} c_k^{\mu\nu} \; ; \; A_{ijk\ell}^{\mu} = \sum_{\nu=1}^{N^{\mu}} K_{ij}^{\mu\nu} c_k^{\mu\nu} c_\ell^{\mu\nu} \; ; \; etc$$

Other structural sums may be derived from these, for example

$$\Xi_{in} = \varepsilon_{ijk} \varepsilon_{\ell mn} \overline{A}_{jlkm}$$

B.2.2 *Other notations*

$\overline{q}$	(over-bar) assembly-average value of the quantity q
q'	(prime) fluctuation of the quantity q
$\hat{q}$	(hat) Fourier-transformed of the quantity q
$q^{\bullet}$	(superscript dot) value of q for a generic particle number
$\dot{q}$	(over-dot) rate of change (time differentiation)
$q^{(mf)}$	mean-field value of the quantity q

Index

Q

R

S